Linked data: a Geographic Perspective

Linked data: a Geographic Perspective

Glen Hart • Catherine Dolbear

CRC Press
Taylor & Francis Group
Boca Raton London New York

CRC Press is an imprint of the
Taylor & Francis Group, an **informa** business

CRC Press
Taylor & Francis Group
6000 Broken Sound Parkway NW, Suite 300
Boca Raton, FL 33487-2742

First issued in paperback 2019

ISBN-13: 978-1-4398-6995-6 (hbk)
ISBN-13: 978-0-367-86654-9 (pbk)

Library of Congress Cataloging-in-Publication Data

Hart, Glen, 1959-
    Linked data : A Geographic Perspective / Glen Hart and Catherine Dolbear.
      p. cm.
    Includes bibliographical references and index.
    ISBN 978-1-4398-6995-6
    1. Geography--Data processing. 2. Geography--Computer network resources.
  3. Semantic Web. I. Dolbear, Catherine, 1976- II. Title.

G70.2.H37 2013
910.285'4678--dc23
                                               2012027810

Visit the Taylor & Francis Web site at
http://www.taylorandfrancis.com

and the CRC Press Web site at
http://www.crcpress.com

What is written without effort is in general read without pleasure.

Samuel Johnson

*We hope we have made enough effort to
bring some pleasure in reading.*

Glen Hart and Catherine Dolbear

Contents

Preface

WHAT THIS BOOK IS ABOUT

This book is a practical guide to implementing solutions based on Linked Data and the Semantic Web that involve Geographic Information (GI). The intended audience is those interested in using GI as part of the Semantic Web, as well as GI professionals wanting to understand the impact of Semantic Web technologies on their work. Much of what we say will also be relevant to anyone interested in publishing Linked Data or preparing ontologies (semantically rich vocabularies to describe data). This preface is not the place to define the Semantic Web, Linked Data, or GI in any detail, but we can briefly set out the terminology here. The Semantic Web is an extension of the Web that enables people and machines to understand the meaning of the data on the Web more easily; Linked Data represents part of the Semantic Web and refers to a way of publishing structured data on the Web and interlinking it together; and Geographic Information is exactly what it says it is—any information with a geographic element. That you are reading this book indicates that you are at least aware of one of these terms. We hope you will forgive us for not expanding these terms further at this time and can wait for the beginning chapters for all to be revealed.

We admit up front that writing this book has been a challenge as it brings together two quite different disciplines—GI and the Semantic Web—and attempts to tell their stories to three different audiences: those with a knowledge of GI but not the Semantic Web; those with a knowledge of the Semantic Web and not GI; and those to whom both these topics are new. Between us we have experience of all three communities, having started as we all do in the last community, and therefore hope that through our combined experiences we will connect to all our readers.

WHY WE WROTE THIS BOOK

When reading a technical book, we believe that from the start it is important that the reader understands the authors' motives for writing it. From the outset, we realized that this book will not make us rich, although we do admit to a degree of vanity. Over and above this understanding, our motives can best be appreciated if the reader knows how we became involved in this field.

We became actively involved in the Semantic Web in 2005 when we were both working within the research group at Ordnance Survey, Great Britain's national mapping agency. A knowledge of GI was therefore deeply embedded in the organzation. At that time, most activity concerned with the Semantic Web was concentrated in universities. Few outside academia had heard of the Semantic Web, the term *Linked Data* was not known at all, and we were fortunate to work within an organization that encouraged the investigation of novel technologies. We recognized early on the value that the Semantic Web could bring to data integration based around GI. We were equally aware that the manner in which large parts of the GI industry had developed had resulted in its overspecialization and insularity. The GI industry had developed standards and tools that were largely unknown outside that community, making it difficult to use as a medium for data exchange or integration. (To be fair to those working with GI, this is a common artefact of many close technical communities.) We felt that the Semantic Web could offer a means of opening up the GI world, so that it would not only be more useful for data integration, but also make it available for more widespread social use. In essence, we saw the potential to make GI more accessible through the application of Semantic Web technology.

We embarked on a series of projects to investigate how to describe GI using Semantic Web technology and how to use this technology to help to integrate Ordnance Survey's data with other data. We rapidly discovered that at that time few people understood both GI and the Semantic Web, and our only route was to learn through experience. Even now, it is apparent to us that too few people really understand both disciplines in depth, and most of these individuals are in academia. In this respect, our position at Ordnance Survey as part of the research group was fortuitous; it provided us with easy access to academia and at the same time grounded us in the practical needs of an organization constantly striving to improve the nature, quality, and applications of the GI it produced. Thus, we believe that we have developed not only a good understanding of GI and the Semantic Web but also practical experience of using these technologies together in an industry context, much of it learned the hard way. So, we wrote this book from a desire to impart the knowledge that we had acquired, such that the use of GI in the Semantic Web would be accelerated and, in its turn, the publication of more GI in this form would grow the Semantic Web.

Another frustration that we had encountered, whether dealing with the Semantic Web or GI, was how difficult it was to find material that explained these concepts in layman's terms. To a very large degree, this is due to the fact that by nature these subjects are very technical and therefore difficult to express in the vernacular. However, we felt that not enough effort was being put into making them more accessible; hence, another reason for writing this book is to attempt to fill this gap and to communicate to an audience that is wider than just the engineer and technician. Therefore, it has been our intention to make the book as readable as possible; we will help you to pick a way through the minefield of jargon and acronyms that obscure all technical subjects. Where we give examples, we will do so using simple English whenever possible, and where fragments of code are presented, they will be clearly annotated and expressed in the most user-friendly notation possible.

HOW TO READ THIS BOOK

You may be very familiar with GI, but not be that knowledgeable about the Semantic Web, or quite the reverse, or you may be learning about both. Equally, you may regard GI as absolutely central to what you are doing, or it may play just a small part, perhaps being no more than the need to show some data on a map background. To meet this challenge, we have attempted to structure the book such that there are introductory chapters for GI, the Semantic Web, and Linked Data before we embark on an explanation about how these can operate together.

The task we have set ourselves is particularly difficult as the Semantic Web and Linked Data are still rapidly developing. Our approach has been to concentrate on those aspects that are most firmly rooted, such as the manner in which data can be represented as Linked Data, and concentrate much less on specific software solutions. Here, it is our hope that you will be able to supplement the durable knowledge that we impart with future knowledge about specific implementation technologies from contemporary publications. We are therefore not supplying a complete solution: What we attempt to do is impart enough insight to enable you to think about GI through the lens of the Semantic Web.

We hope you will find this book not only informative but also an enjoyable read.

Glen Hart and Catherine Dolbear

About the Authors

Glen Hart currently leads the Research group at Ordnance Survey, Great Britain's national mapping agency. He has degrees in computer science and natural science. Glen began his career working for the U.K. defense industry before becoming a specialist in engineering software. In the early 1990s, Glen joined Ordnance Survey, initially working on corporate spatial data models and spatial data strategy before joining the research group. Over the years, he has become more of a geographer whilst maintaining a keen interest in computer science. Initially, Glen's research interests concentrated on data integration, but this broadened to include land cover and land use classification, vernacular geography, and crowd sourcing. It was his initial interest in data integration that made him both aware of and interested in the Semantic Web. Glen first began to investigate the Semantic Web in 2002, initially working to understand the relevance to data integration and then how to construct ontological descriptions of geographical objects. The aim was to investigate the usefulness of ontologies in data integration. Mirroring his interest in vernacular geography (how people understand and refer to the landscape around them as opposed to official views), Glen was also determined to develop ways to make ontologies more accessible to domain experts, those people expert in a subject area but for whom ontologies expressed in formal logic were totally opaque. This resulted in the development of a syntax for the Web ontology language OWL, called Rabbit, that expresses OWL sentences using controlled natural language. More recently, Glen has been involved in the publication of Ordnance Survey's geographic information as Linked Data. Glen sees the publication of linked data as an enabler towards more efficient and accurate data integration, and his research continues along these lines.

Dr. Catherine Dolbear is currently a linked data architect at Oxford University Press (OUP), working on strategies for linking content across online academic and journal products within OUP's discoverability program. She has a degree in electronic engineering from Imperial College, London, and a DPhil in information engineering from the University of Oxford. Catherine is the author of several papers in areas of geographical ontology development, semantic data integration, and information filtering using structured data; a previous cochair of the OWL Experiences and Directions Workshop; and cofounder, along with Glen Hart, of the international workshop series "Terra Cognita" on geospatial semantics.

Catherine first encountered the Semantic Web during her doctoral research on personalized Web information filtering and expanded this interest to include geographical information while she was leading the geosemantics research team at Ordnance Survey. She has experience in knowledge elicitation and user evaluations and applied this in her work with domain experts to capture their knowledge in ontologies. Catherine has worked on the development of several ontologies in the geographic domain, such as buildings, places, hydrology, and spatial relations as well as the reuse of these in application areas such as flood defense and risk management.

Her research at Ordnance Survey was focused on opening up geographic data for multiple uses by using semantics to repurpose relational data. More recently, she has been interested in moving semantic technologies into mainstream publishing and tackling the scale, quality, and workflow requirements thereof. She is interested in how people think about the world, what is important to them, and how to discover knowledge by forging new links between information.

And those who helped

Books are rarely just written by the authors alone and this book is no exeception. We have taken advice and learnt from many. In particular we mention and thank Dr John Goodwin with whom we have both worked closely and have benefited from his knowledge.

1 A Gentle Beginning

1.1 WHAT THIS BOOK IS ABOUT AND WHO IT IS FOR

The majority of people have not heard of terms such as the Semantic Web, Linked Data, and Geographic Information (GI); these terms are at the very most known of only vaguely. GI is the "old boy," having been around as a term for at least thirty and perhaps forty years, but has existed within a relatively small community of experts, a growing community but nonetheless small compared to the total size of the "information community" in general. As its name implies, it refers to any information that has a geographic component. The terms *Semantic Web* and *Linked Data* are the "new kids on the block." The Semantic Web is an extension of the Web, which enables people and machines to understand the meaning of the data on the Web more easily. The idea of the Semantic Web was conceived in the late 1990s and as a whole has remained largely in the academic world since that time. Linked Data is a component of the Semantic Web and concerns the manner in which data is structured, interrelated, and published on the Web; as a named entity, it has only been around since the mid-2000s.

The use of GI has been well established within the information technology (IT) community for a long time and is becoming increasingly important as a means to enable geographic analysis and as an aid to integrate data. Linked Data has been growing rapidly since 2007, resulting in the emergence of the Linked Data Web, a part of the broader World Wide Web, but where the focus is on data rather than documents. The Semantic Web as a whole continues to grow much more slowly, in part due its technical complexity and in part due to the lack of data in the right format—the growth of the Linked Data Web is overcoming this last impedance.

The purpose of this book is to provide an explanation of the Semantic Web and Linked Data from the perspective of GI, and from a different angle, it shows how GI can be represented as Linked Data within the Semantic Web. How you view the book will depend on who you are and what knowledge of these topics you already have. The readership is intended to be quite varied. You may be someone who knows about GI but little about the Semantic Web or Linked Data; conversely, you may know something of the Semantic Web and Linked Data but little of dealing with GI; and of course you may wish to find out more about both. Even if your primary aim is to find out about the Semantic Web and Linked Data and you are not that interested in GI, the book still has much to offer as the examples we give are not unique to GI.

We also intend this book to be very much an introduction and have tried to write it in a manner that makes it accessible both to those with a technical mind for IT and to those who may not be so technically knowledgeable of IT but still have to be aware of the nature and potential for these topics, perhaps as managers, business leaders, or end users.

The book does not attempt to provide a very detailed implementation guide, however: It is not a hard-core coders' manual. Instead, it is an introduction to the topic with an emphasis on getting the approach right at the design level. Nonetheless, the book introduces all the main technologies and provides reasonably detailed descriptions of the languages and syntaxes that are associated with the Semantic Web and Linked Data in particular.

The terms *Semantic Web* and *Linked Data* lead one to believe that these technologies are only for the Web, but this is not the case; more and more people are also turning the technologies inward and using them within their organizations as well as to publish data externally on the Web. These technologies are more than anything about data integration; there is a big clue to this in the term *Linked Data*. GI also has an important role to play in integrating data. As we discuss further in the book, the very characteristics that make these technologies suitable for data integration also make them good at adapting to change—something that all organizations face and with which traditional ITs struggle to cope. The book is therefore also aimed at those for whom better ways to structure their organization's data are also a driving force, as well as for those whose aim is to publish their data on the Web.

By the end of the book, you will have a good understanding of the nature of the Semantic Web, Linked Data, and GI. You will be able to think about how to model information, especially GI, in a way suitable for publication on the Linked Data Web and which is semantically described. Technically, the book will have provided an overview of the key languages that you will need to master (RDF, the Resource Description Framework; OWL, Web Ontology Language; and SPARQL—the recursive acronym for SPARQL Protocol and RFD Query Language), and you will understand the process required to publish your data. Last, you will have been introduced to some of the tools required to store, publish, interlink, and query Linked Data.

1.2 GEOGRAPHY AND THE SEMANTIC WEB

Before descending into too much detail, let us start by introducing the terms *Geographic Information*, *Semantic Web*, and *Linked Data*.

1.2.1 GEOGRAPHIC INFORMATION

Put simply, GI is any data that has a geographic aspect, although in many cases the geographic component may be quite small or viewed as only peripheral to the main focus. To those whose first encounter with GI has been through location-based applications such as Google Maps, it may not be immediately apparent why GI has traditionally been seen as a distinct class of information. Special systems to handle geographic systems have been around since the early 1960s. At first, these were all bespoke systems, as indeed were most computer applications of the time, although by the late 1960s to early 1970s commercial off-the-shelf Geographic Information Systems (GIS) were becoming available. The reason for the emergence of this distinct class of software GI has been attributed to the need to perform specialist computations involving geometry. For example, the one-dimensional indexes suitable for mainstream databases had to be supplemented with two-dimensional spatial indexes within GIS. GI has matured in conjunction with the development of GIS, and there have been spatial extensions to mainstream database products such as Oracle, DB2, SQL Server, and MySQL. Most recently, the growth and usage of GI have been affected by the implementation of Web-based tools such as Google Earth and the creation of GI resources by volunteers, an example being OpenStreetMap, a Web-based cartographic resource.

The use of GI has grown enormously, but many users will be largely unaware of the role that GI plays in their activities. This is because GI is very rarely an end in itself; rather, it normally forms the canvas on which the subject of interest is painted. Thus, GI is present in a wide range of subject areas—ecology, land administration, town planning, insurance, marketing, and so on—but rarely as the main focus. For something that is so obviously visual, it is therefore somewhat ironic that it is so often unseen. Since there is a geographic context to so many subject areas, GI has an important role to play in linking and combining datasets through shared location. As yet, this potential for data integration based on GI is far from fully realized. In part this is because the data itself is often not well organized; in part, it is because the technology has not been there to aid the process.

1.2.2 THE SEMANTIC WEB

The World Wide Web had not yet been born during GI's formative years, while the Semantic Web dates just to 1998 (Berners-Lee, 1998c) and so is barely in its teens. Here in this introduction, we do not go into detailed explanations about the Semantic Web; that is for further in the book. For the moment, it is sufficient to say that the Semantic Web provides a means to describe, query, and reason over both data and Web content using a combination of established Web technologies such as HTTP (Hypertext Transfer Protocol), a universal data structure (RDF, the Resource Description Framework), a means to query that data (SPARQL), and a means to semantically describe and annotate the data using ontologies (formal machine-readable descriptions), usually expressed in RDFS (RDF Schema) or OWL (Web Ontology Language).

Linked Data, an integral and essential part of the Semantic Web, refers to a way of publishing structured data on the Web (using RDF) and interlinking it. The notion of Linked Data itself can be traced to the very origins of the Semantic Web; however, the term *Linked Data* was only established later. Linked Data has been seen as separate from the Semantic Web or even as the only part of the Semantic Web that works. However, to us Linked Data is an integral part of the Semantic Web, and at the time of writing, it was the most actively growing and developing element. Although at times we do treat Linked Data on its own, if we do so it will be when Linked Data is the particular focus, and when we do so, we always treat it *as a part of* the Semantic Web. If we refer to the Semantic Web, we always do so in an inclusive manner with respect to Linked Data.

1.3   GI IN THE SEMANTIC WEB

We believe that GI has certain aspects that require more careful consideration when using it within the Semantic Web, but there is little if anything that is truly unique about GI. Perhaps the things that have made it special in the past, such as its concentration on geometry, will be less important in the Semantic Web era, helping to make GI more accessible. We argue that it is the ability of GI to facilitate the integration of other data that is of real interest in the context of the Semantic Web. The Semantic Web is *about reasoning over data and integrating data*; it therefore makes sense that GI, as an enabler for integration, and the Semantic Web are natural bedfellows.

There are peculiarities about GI that present particular challenges with respect to the Semantic Web. Geometry is certainly one of them, and there are others, such as the application of topologic networks and the vagueness that exists in geographic categories. For example, what makes a river a river and not a stream? Again, it is worth emphasizing that these peculiarities are not unique to geography. And, it is not always possible to provide complete solutions to these problems. In these cases, we therefore suggest how best to manage the problems, compromising an unobtainable ideal for the best practical solution permitted by contemporary technologies and information theory. It is the intention of this book to cast light on the main challenges of using GI within the Semantic Web and explain the best practice for its use, as bounded by the constraints of current technology and knowledge.

1.4 EXAMPLES

To explain the application of Semantic Web and Linked Data techniques to GI, we make copious use of examples. In places we have done so in a way that shows how an organization may develop its solutions, and to do this we have decided to create an imaginary country—the island state of Merea—along with a number of organizations within that state. Principal among these is Merea Maps, an imaginary national mapping agency. Merea Maps appears very prominently partly because we both have worked for such an organization and are thus familiar with the challenges faced by mapping agencies. But more important, the work of mapping agencies provides a rich selection of examples in their task of not simply producing mapping but of creating digital representations of the world where we live. In doing so, they encounter a multitude of interesting challenges and face a world where precision is not always possible and uncertainty is faced on a regular basis. The data they produce is used as a fundamental component of many other applications; therefore, we can also follow the story of this data, from how it is collected and represented by Merea Maps through how it can be combined with other data by a third party to address specific problems.

Merea Maps is intended to be a mirror of any modern mapping agency. For those not familiar with mapping agencies in the twenty-first century, it is worth emphasizing that these organizations are not simply responsible for providing traditional paper mapping. Their main role is to supply high-quality digital information about the topography related to their area of interest. In many countries, this responsibility may also extend to hydrographic mapping and maintaining a land cadastre (a registry of who owns what land). However, for the sake of simplicity Merea Maps concerns itself with inland topographic mapping. The digital products produced by such agencies can be roughly divided into four types:

- Digital topographic maps, the modern equivalent of a paper map.
- Gazetteers, geocoded lists of places and addresses. These are like indexes to the landscape: They enable services to locate the position on Earth's surface of a place or an address; conversely, someone can use them to find the place or address at a particular location.
- Terrain models. These are digital models of Earth's surface.
- Photography and other sensor data such as Lidar–typically derived from either aircraft or satellites.

These data form the basis for other organizations to build on by either adding additional information or using the data to perform geographic analysis related to their business, such as a retail chain using GI to work out ideal store locations or an insurance company working out insurance risk for areas prone to flooding. Thus, an important element of any modern mapping organization is to deliver data to its customers in a form that is easy for the customers to accept and use. This need in particular is why mapping organizations are one type of many organizations today looking at Linked Data and the Semantic Web as better ways to perform their role.

1.5 CONVENTIONS USED IN THE BOOK

The later chapters of this book contain many examples, and to make these examples easier to understand, we have used a number of conventions to represent the nature of elements of these examples.

In diagrams, we have adopted the convention that concepts or classes (abstract categories that real things can be placed into such as car, building, and river) as rectangular boxes with rounded edges and instances or individuals (i.e., actual things such as your car, the White House in Washington, and the Amazon River) as ellipses. Relationships between classes and individuals are shown using directed arrows. These conventions are shown in Figure 1.1.

So, using these conventions we can say unambiguously that the White House (an individual) is a Building (a class) and so on. The book also illustrates points using "code." The code represents an example in one of a number of different computer languages and syntaxes: RDF and RDFS using RDF/XML (eXtensible Markup Language) and Turtle syntaxes and OWL using OWL/XML, Manchester Syntax, and Rabbit. We have used different syntaxes because a number of different syntaxes currently exist, and there is no ideal for showing all the examples consistently: RDF/XML and OWL/XML are able to express all the examples but are verbose and difficult to understand, whereas Turtle, Manchester Syntax, and Rabbit are much easier to understand but cannot express all the examples. As a general principle, we have usually chosen to use the most understandable syntax. We have also adopted the principle of showing Manchester Syntax and Rabbit side by side as although Rabbit

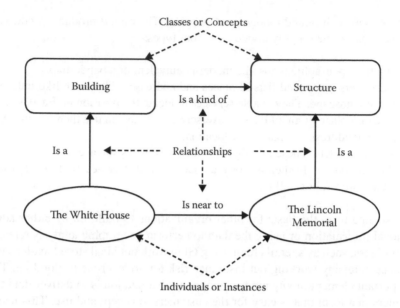

FIGURE 1.1 Diagramming conventions.

is more understandable, Manchester Syntax is more widely used and has much better tool support. We have also depicted all code examples using a "computer-style" pseudocode to emphasize that it is code. Code examples are as follows:

RDF/XML:

```
1   <?xml version = "1.0" encoding = "UTF-8"?>
2   <rdf:RDF
3       xmlns:rdf = "http://www.w3.org/1999/02/22-rdf-syntax-ns#"
4       xmlns:mereaMaps = "http://mereamaps.gov.me/placesOfInterest/">
5   <rdf:Description
6       rdf:about = "http://mereamaps.gov.me/placesOfInterest/0012">
7       <rdf:type rdf:resource = "http://mereamaps.gov.me/
placesOfInterest/Pub"/>
8       <mereaMaps:has_name>The Isis Tavern</mereaMaps:has_name>
9       <mereaMaps:has_longitude>-1.241712</mereaMaps:has_longitude>
10  </rdf:Description>
11  </rdf:RDF>
```

Turtle:

```
1   @prefix rdf: <http://www.w3.org/1999/02/22-rdf-syntax-ns#>.
2   @prefix mereaMaps: <http://mereamaps.gov.me/placesOfInterest/>.
3
4   mereaMaps:0012 mereaMaps:has_name "The Isis Tavern".
5   mereaMaps:0012 mereaMaps:has_longitude "-1.241712".
```

The RDF/XML and Turtle examples both include line numbering, which is not part of the syntax but has been added to aid explanation.

Rabbit and Manchester Syntax side by side:

Rabbit	Manchester Syntax

`Every Farm has part a Building.`
`Every Farm has part a Field.`
	`Class: Farm`
	`SubClassOf: hasPart some Building,`
	`          hasPart some Field`

1.6   STRUCTURE OF THE BOOK

Chapters 2 to 4 form a foundation for the rest of the book. Chapters 2 and 3, respectively, introduce the reader to the Semantic Web and Linked Data and to GI. For those not familiar with the Semantic Web or Linked Data, Chapter 2 is important as it introduces not only these technologies but also something called the open world assumption, a way of thinking about how information is interpreted on the Web. Chapter 4 then brings together GI and the Semantic Web and importantly applies the open world assumption to GI.

Chapters 5 to 8 focus on Linked Data and GI. Chapter 5 introduces RDF, the language used to represent Linked Data, and Chapter 6 then demonstrates how GI can be represented using RDF. Chapters 7 and 8 look at how Linked Data is published on the Web and how it can be linked to other data.

Chapters 9 and 10 then switch from Linked Data to the wider Semantic Web through the description of information-using ontologies (formal descriptions that are machine interpretable). Chapter 9 introduces the Web ontology language OWL, and Chapter 10 shows how OWL ontologies can be developed for GI and demonstrates how these ontologies can be used to infer more about the content of Linked Data than is simply represented in the data and how the content can be used to aid data integration.

Chapter 11 provides a conclusion and summary and makes some predictions for the future, providing a natural end to the reader's journey as all good books should.

1.7   A LAST THOUGHT ABOUT HOW TO READ THIS BOOK

Before embarking on the main body of this book, it is worth considering what this book tries to accomplish. Much of the content is inevitably about technology, whether the technology behind GI or that associated with the Semantic Web and Linked Data, so it is not unreasonable that you will spend much of your time trying to understand the technology. However, the book also presents arguments that may require you to think differently about how you approach the representation and interpretation of information. Geographers who have not encountered Linked Data and the Semantic Web may initially find it difficult to understand the basic mathematical logic that underpins these technologies or the nature of the open world assumption.

Those more comfortable in these areas may be amused by the lack of certainty that inhabits the world of geography, whether this concerns the lack of boundaries between objects or the lack of clear differentiators in terms of different classes of things: When is a river a river and not a stream? In either case, the main aim of this book is to provide the reader with a sufficient understanding of these topics to enable the confidence to organize and publish GI on the Semantic Web.

2 Linked Data and the Semantic Web

2.1 INTRODUCTION

This chapter gives an overview of the Semantic Web mission to add meaning to the World Wide Web, and it introduces the main concepts involved. To set the topic in context, we cover the early history of the Semantic Web as it developed from the World Wide Web. Its emergence was rooted in the need to provide more meaningful search results based on a real understanding of what the Web page was about: its *semantics*. In the early days, there was a tendency to focus on modeling high-level abstract concepts in an attempt to create generalized models of the world, often using first- or higher-order expressive logics. These attempts to "model the whole world" ran into problems of scope creep and complexity, so later work focused on the development and use of tractable subsets of first-order logic and ontologies that were designed with a specific purpose in mind.

This chapter explains the main benefits of the Semantic Web, including its use in data integration and repurposing, classification, and control. We also explain the relationship between the Semantic Web and Linked Data, preparing the reader for further chapters that cover the process of publishing information as Linked Data and authoring Semantic Web domain descriptions known as ontologies.

2.2 FROM A WEB OF DOCUMENTS TO A WEB OF KNOWLEDGE

The fundamental unit of the World Wide Web is the document. Each Web page is basically a document that is connected to other documents via hyperlinks. A user searches for information or finds the answer to a question by reading a Web page that they hope contains the information sought. This Web page will have been retrieved by a search engine, which ranks the relevance of the pages by analyzing the links between them. One example of this is Google's PageRank algorithm (Brin and Page, 1998), which measures the relative importance of a page within the set of all Web pages. This importance measure depends first on the number of links to the page and second on the PageRanks of the pages that display those incoming hyperlinks. Therefore, a page that is pointed to by many pages that themselves have a high PageRank will also earn a high rank. A document's PageRank is the probability that a Web user clicking on links at random will arrive at that document. What this means, then, is that there is no understanding within the search engine about what knowledge that Web page contains, and algorithms based on link analysis cannot distinguish between synonyms. For example, a search for the word *bank* will return information about either a financial bank or a river bank, whichever is the most popular and linked to.

Aside from text documents, information can also be accessed from the Web via Web applications or services; from Web-based e-mail to booking an airline ticket, these allow the user to send and receive information from the Web site, according to the Web site's own particular data structures and capabilities. The user's data (e.g., e-mail address or flight details) will be stored in a database, which is accessed by the Web server. Many Web sites now allow other software developers to access the capabilities of the service, and hence the data that sits behind their application via an Application Programming Interface, or API. For example, the Facebook API allows developers to add a "Like" button to their Web site or enables users to log in to the third-party Web site using their Facebook log-in details. The API in effect specifies what queries can be asked of the service and how the response from the service should be interpreted.

While older methods of describing Web services using CORBA (Common Object Request Broker Architecture), RPC (Remote Procedure Call), or SOAP (Simple Object Access Protocol) describe what the Web service can *do*, a more recent development in Web service design is the REST (Representational State Transfer) architecture, which concentrates on describing the *resources that the service provides.* RESTful applications use HTTP (Hypertext Transfer Protocol) requests to post (create or update) data, delete data, or make queries (i.e., read the data). A REST service can respond with the data formatted in CSV (comma-separated value) files, JSON (Java Script Object Notation), or XML, and the service can be documented (that is, specifying what data is available from the service and what queries can be asked) using version 2 of WSDL (the Web Service Description Language) or can be simply described using a human-readable HTML file. However, although the data access mechanism has been greatly simplified using REST, the *structure* of the data still remains proprietary to each Web service, and if the structure of the XML output were to be changed, say, it would break any client that currently consumes the data. Furthermore, there is no way of indicating links *out* of the individual dataset, that is, there is nothing analogous to an HTML link, which could signpost where the authors of the dataset believe it might overlap with other datasets.

So, if a programmer wants to write an application that uses data from several APIs (e.g., to plot where all the programmer's Facebook friends live on Google Maps), the programmer would have to understand the structure of the data returned by each separate API and discover all the relevant links between the two datasets. For a well-coded API, this may be relatively straightforward, but it cannot be done automatically, and it cannot be done in the general case. However, if the data were published in a structured format that was commonly understood, and if there was a description of the meaning of the data, this would make each dataset not only understandable but also discoverable: This is what Linked Data sets out to do. This is particularly important for Geographic Information (GI), because GI is so often the common element between datasets. In effect, publishing Linked Data provides an alternative to publishing an API.

The idea behind the Semantic Web is to describe the meaning of the information that is published on the Web to allow search and retrieval based on a deeper understanding of the knowledge contained therein. The Semantic Web adds structure to Web-accessible resources so that they are not only human readable but also can be understood by computational agents. Instead of text, which cannot be processed by a computer without analysis by complex natural language processing algorithms, information is published on the Semantic Web in a structured format that provides a description of what that information is about. This means that the fundamental unit of the Semantic Web is an item of data, or a fact, rather than a document. These facts are brought together to describe things in the real world. Each fact has a basic structure known as a triple as it comprises three elements: a subject, a predicate, and an object. These may either be categories of things, such as "River," individual things such as the "River Thames," or just a value. Inevitably, a thing is described in terms of other things, and this introduces relationships between things, which parallels the concept of hyperlinks between documents in the traditional Web. For example, the relationship "flows through" can link the individuals "River Thames" and "London." Triples can be interlinked through shared subjects or objects as shown in the following example:

Subject	Predicate	Object
Des Moines	is part of	Iowa.
Iowa	is part of	The United States of America.
The United States of America	has population	313000000

So that it becomes possible to navigate between facts, in the example we are able to go from the fact that Des Moines is in Iowa, to learn that Iowa is the in the United States of America, and then to find out that the United States has a population of 313,000,000.

Many types of predicate or relationship are possible between both categories and individuals, such as hierarchical ("is a kind of"), mereological ("is a part of"), or spatial ("is adjacent to"), and to specify that an individual is a member of a category ("is an instance of"). In fact, the author of a Semantic Web description can choose any relationship the author wishes, thus adding rich meaning to the data on the Semantic Web. What is particularly interesting is that the item of data may reside in one document, or dataset, but be linked to another data item in a different dataset, by any relationship, but particularly the equivalence relationship "sameAs," thus instigating a process of data integration.

Collections of statements about related things in a particular subject area or domain can be grouped together in what is known as an ontology. An ontology is therefore more than just a vocabulary, which specifies which terms can be used; or a taxonomy, which classifies instances into classes. It is a knowledge representation—a specification of a number of concepts in a domain that are described through the relationships between them. Some commentators (e.g., in the Semantic Web OWL standard) (Smith, Welty, and McGuinness, 2004)) point to a further characteristic of ontologies: reasoning. Statements encoded in OWL can be processed by software that can reason over them (known as "inference engines," "rules engines," or simply "reasoners"). These reasoners infer logical consequences from a set of axioms or facts. For example, if we state that "Every Estuary flows into a Sea"[1] and that the "Thames Estuary is an Estuary," we can then derive the additional information that the "Thames Estuary flows into a Sea." Contrasting this phenomenon to the ordinary

Web, it is as if new links were being created between nodes on the Web. The facility of reasoning is the main differentiator between an ontology and other data structures used on the Web, such as XML schema.

2.3 EARLY HISTORY AND THE DEVELOPMENT OF THE SEMANTIC WEB

The original proposal for the World Wide Web (Berners-Lee, 1989) already included a discussion of how a node on the Web could represent any object, person, project, or concept and not necessarily just a document. The arrows that linked the nodes could mean "depends on," "is part of," "made," "refers to," "uses," "is an example of," and so on, adding a meaning beyond simply "connected to" that the original World Wide Web began to employ. In Tim Berners-Lee's vision, this would allow "links between nodes in different databases."

Berners-Lee expanded on this idea in a well-known *Scientific American article*, where he, Jim Hendler, and Ora Lassila coined the term *Semantic Web* (Berners-Lee, Hendler, and Lassila, 2001) as "an extension of the current [Web] in which information is given well-defined meaning, better enabling computers and people to work in cooperation" p. 29. It pointed out the weakness of many knowledge representation systems at the time: They were centralized. This meant that even if the data they contained could be converted from one format to the other, the logic, or as they put it, "the means to use rules to make inferences, choose courses of action and answer questions," could not be easily reused by other systems. Furthermore, these systems could not be scaled easily, especially not up to the size of an "ecosystem" as large as the World Wide Web.

The Semantic Web, according to Berners-Lee et al., would employ a standard knowledge representation language and simple rule declaration language, in which information on Web pages could be encoded. This structured information could be used to improve the accuracy of a Web search. The search engine would look only for the pages that contained markup referring to the precise concept, rather than relying on vague keywords. More complex applications could link the structured data on the Web page to an ontology that included inference rules and return results based on the inferred information. The *Scientific American* article, and other early descriptions of the Semantic Web, primarily concentrated on how to bring meaning to the information already published in Web pages. This spurred the development of many ontologies, but they tended to suffer from problems of centralization and "scope creep": the misplaced and futile desire to capture all possible knowledge in one place.[2] There was also the "cold-start" problem: how to convince people to mark up pages that already existed without the clear benefit that would come when the technique was widely used. It was only later that the focus of Semantic Web research shifted to accessing data already structured in relational databases.

The early years of the twenty-first century saw the development of new ontology editors, such as Protégé at Stanford University (Noy and McGuinness, 2001), and efforts to standardize the languages that were to become the main tools of the Semantic Web: the Resource Description Framework (RDF) (Manola and Miller,

2004) and the Web Ontology Language OWL. There was also significant effort in the development of reasoners to reason over OWL such as Pellet (Sirin et al., 2007) and Racer (Haarslev and Möller, 2001), as well as the development of query languages to interrogate the data encoded in RDF, such as SPARQL (Prud'hommeaux and Seaborne, 2008) and the less-used RQL (Karvounarakis et al., 2002), RDQL (Seaborne, 2004), and SeRQL (Broekstra and Kampman, 2003). There were also many new ontologies appearing, representing the knowledge in domains such as bioinformatics, for example, the Gene Ontology,[3] law (Hoekstra et al., 2007), and geography.[4] This led to many interesting results in knowledge representation, particularly the rise of ontology design patterns (Gangemi, 2005).

An ontology design pattern is a "reusable successful solution to a recurrent modelling problem"[5] that can help authors to communicate their knowledge within the confines of their chosen ontology language. However, at this time there was also something of a struggle between the various ontology languages for dominance. Suffice it to say for now that there was the choice between more complex and expressive languages (such as the Knowledge Interchange Format or KIF [Genesereth and Fikes, 1992], which used first-order predicate calculus; or CYCL [Cycorp, 2002], which used higher orders of logic); simpler languages such as the precursor to OWL called DAML+OIL (Connolly et al., 2001) and RDFS (Brickley and Guha, 2004) used to describe RDF vocabularies; or ones based on Description Logics, such as the OWL variant OWL-DL, which balanced expressivity against complexity. That is, the authors of OWL-DL tried to strike a balance between expressing knowledge of the domain accurately and guaranteeing, within a finite time, that the reasoner operating on the language would find all the correct answers to the query posed.

Another research theme at this time was the development of so-called upper ontologies such as the early UML-based knowledge model CommonKADS (Schreiber et al., 1999) or more recently SUMO (Pease, Niles, and Li, 2002) and DOLCE (Gangemi et al., 2002). These ontologies, also known as "top-level" or "foundation" ontologies, describe concepts at a very general level to lay a "foundation" for knowledge above any individual domain. An ontology that describes knowledge about a specific domain can then inherit from the upper ontology. That is, all concepts in the domain ontology can be expressed as subclasses of concepts in the upper ontology. This then assists interoperability between domain ontologies based on the same upper ontology. For example, the DOLCE ontology specifies concepts like *endurant* (a concept that can change over time while remaining the same concept, such as a person) and *perdurant* (a concept that occurs for a limited time, such as an event or process). An ontology describing a specific domain would then inherit from DOLCE; for example, a river would be a kind of endurant and a flood would be a kind of perdurant. A second ontology, say on habitats, that is also based on DOLCE might contain a concept like freshwater habitat, which because it is also a kind of endurant could then be related to the river concept in the first ontology as they share a common ancestral concept. Clearly, this example is very simplified, and many more layers of granularity would necessarily be included in the ontology to fully describe the domain.

There are, however, some drawbacks to using upper ontologies, not least because it can be very difficult for an expert in a particular domain such as GI to understand exactly which of the oddly termed classifications to assign to their concepts. Should a

County be classed as a Physical Region or a Political Geographic Object? Is a flood an endurant or a perdurant? It depends on your point of view. These quandaries become even more apparent when confronted with terms like "Non-Agentive Social Object" or "Abstract." The domain expert can spend more time puzzling over where to place a domain concept in the upper ontology hierarchy than actually building his or her own ontology. Furthermore, upper ontologies are inherently based on the principle of centralization: If all our domain ontologies are based on the same upper ontology, it will be easy to integrate them. However, as we have mentioned, the Web is inherently decentralized, and there is more than one upper ontology to choose from, so in this book we would not particularly recommend that the GI domain expert should use an upper ontology as a base when building an ontology. Upper ontologies are useful to study as they present good knowledge modeling discipline when considering the more difficult to model concepts like matter ("Sand" or "Rock") or the difference between Constitution and Identity (e.g., a stream is a body of water, but even if it dries up during summer it is still a stream).

In the same vein as upper ontologies, in the early days of the Semantic Web several large-scale domain ontologies went into development, for example, the very successful GALEN ontology for the biomedical domain (Rector, Rogers, and Pole, 1996). These ontologies tended to consist of hundreds of thousands of concepts, usually in a scientific field, and were authored by committee. In the GI domain, ontological development was influenced by philosophy and a desire to develop a "complete ontology of the geospatial world [that] would need to comprehend not only the common-sense world of primary theory but also the field-based ontologies that are used to model runoff and erosion" (Smith and Mark, 2003). Authoring ontologies is discussed further in Chapter 10, but for now, we can just note that, almost ten years later, this holy grail of a single complete ontology of the geospatial world does not exist, and there is a persuasive argument to say that one monolithic ontology is not necessary or obtainable. Instead, smaller, more agile, rapidly developed ontologies describing specific domains within geography for specific purposes (e.g., the Ordnance Survey Spatial Relations Ontology[6]) have been more successfully deployed.

It was perhaps the concentration of the Semantic Web research community on developing large ontologies and optimizing ever-more-complex reasoners that led some to rethink the direction in which the Semantic Web was heading. The Linked Data movement started with a return to the drawing board to concentrate on exposing data to the Web that was hidden in proprietary databases, structured in myriad ways. To do this, they recommended structuring data in a standard format: RDF, which could also be used to specify links into and out of each dataset.

As described in Heath and Bizer's book on the subject (Heath and Bizer, 2011), "Linked Data provides a publishing paradigm in which not only documents, but also data, can be a first class citizen of the Web, thereby enabling the extension of the Web with a global data space based on open standards—the Web of Data." This means, therefore, that a GI specialist wishing to share data on the Web could publish it as RDF and include links to other datasets to allow the data to be discoverable by Web crawlers. Since geography is so often used as a backdrop to other information sources, it is frequently used in the Web of Data to link together other RDF datasets that have a geographic element. This is apparent in the picture of the

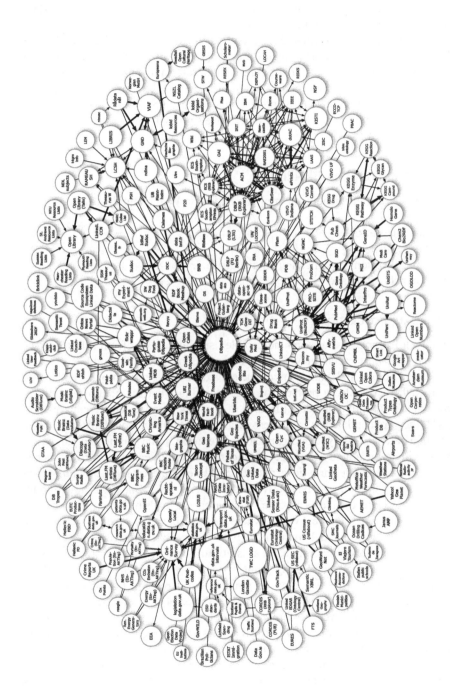

FIGURE 2.1 The Web of Linked Data as of September 2011.

Linked Data Web known as the Linked Data Cloud (Figure 2.1), where GI-based datasets like GeoNames[7] and LinkedGeoData (Hellman, Auer, and Lehmann, 2009) are often more tightly interlinked and act as intermediaries between other datasets.[8]

The technical details of Linked Data are discussed further in Chapters 7 and 8, including a step-by-step guide to publishing geographically based Linked Data. For now, it is sufficient to say that Linked Data concentrates much more on simple structures, data discovery, and reuse compared to the focus on expressing and reasoning over knowledge that characterizes other ontology-based Semantic Web initiatives. Both aspects of the Semantic Web, however, have significant benefits for the use and management of GI.

2.4 SEMANTIC WEB BENEFITS

There are several clear benefits of applying Semantic Web technologies to GI:

1. Data integration
2. Data repurposing
3. Data collection, classification, and quality control
4. Data publishing and discovery

2.4.1 DATA INTEGRATION

Currently, it is expensive to combine complex spatial datasets from different databases and difficult to accurately conflate categories of items. The problem is more than a syntactic one and cannot merely be solved by adherence to one standard file format. Rather, it is also a problem of understanding what the data means: Is one organization's definition of a field (the area of land) the same as a different organization's? If they are conflated as equivalent entities, what impact would that have on the results of a database query? An organization interested in the selling and purchasing of fields will define a field as a parcel of land spatially delineated by barriers such as fences or ditches, while for a government department interested in providing farm subsidies, the definition may be subtly different, with the delineation limited to the extent of crops or accessible pasture. In some cases, the fields referred to may be genuinely the same field in the real world, where for example a field used for pasture has exactly the same location and boundary in both views. But, differences are also possible; as an example, consider a field that is not fully planted with crops: The extent of the field from a land-purchasing point of view will be an area of land defined by its physical boundaries, whereas the government department could consider the field to be the area of land covered by crops. Conversely, different worldviews can classify the same things very differently. A zoologist will define a particular fish in terms of its species and ultimately relate that back to the tree of life as described by a taxonomic system starting with domain at the top, moving through kingdom all the way to species. Seat that same zoologist and the same fish in a restaurant and worldviews change dramatically; now, the fish is classified within "Seafood" and the categorization is entirely different. Both classification systems are equally valid but under very different circumstances. Semantic Web technologies can reduce costs and improve the accuracy of integration by making these semantic differences explicit in ontologies, so the impact on a user's application can be seen. Furthermore, the links and overlaps between the two datasets can be specified exactly, in a more detailed fashion than just making the two categories of Field equivalent, so one query can be carried out on the combination of the two datasets and achieve more accurate results.

2.4.2 Data Repurposing

A second benefit of semantic technologies is in the reuse of data for purposes other than the reason for which it was originally collected. For example, a mapping company might collect information about the spatial extents of objects such as sluice gates and weirs to produce cartographic maps and provide spatial mapping data. An organization reusing that data will have its own tasks to carry out and may need to interpret the terminology differently. For example, the organization tasked with flood risk management would have a need to identify all the possible flood defenses in an area. A simple ontology can link the two views on the world by specifying how the concepts used in one can be related to the concepts used by the other. In our simple example, the statements: "A Sluice Gate is a kind of Flood Defense" and "A Weir is a kind of Flood Defense" enable the topographic objects Sluice gate and Weir to be connected to the world of flood defense. This in turn enables the Flood Defense agency to query the mapping data for flood defenses, even though "flood defense" was not a term explicit in map ontology data, and the mapping company could not have predicted in advance during its data capture process all the possible categories that its customers might in the future want to identify in its data.

2.4.3 Data Collection, Classification, and Quality Control

Semantic technologies benefit the internal business processes of organizations that collect or publish their own content. An ontology provides a very explicit data specification, which assists surveyors, other GI data collectors, or automatic collection mechanisms in their task of identifying the information that needs to be added to or modified in the database and assists with quality control. For example, a surveyor who comes across a small watercourse might need to specify whether it was a bourn or a stream. Many cases might be borderline, and an ontology can help with ensuring data quality. For those designing the data specification, an ontological approach provides a logical framework for addressing questions like, "What categories of object do I need to collect information on?" "How do those categories relate to each other?" "What instructions should I give to the Data Collection department on how to identify and differentiate between Category A and Category B?"

The process of constructing an ontology to describe a data specification also helps the organization clarify to itself exactly why it is capturing the data and whether it is really needed. For example, a common mistake in traditional classification systems is to confuse or conflate different axes of categorization. It is very common for land use classification (the purpose or function of the land) to be mixed up with a description of what the object is, so that a land use classification might be Football Stadium (what something is) rather than its use or purpose: playing football. Similarly, a building's spatial extent or "footprint" is not the same as the building itself. So, one cannot assign a use to the footprint. Instead, best modeling practice is to define the building as the primary object and assign it both a footprint and a use. One might argue that this is all far too much detail, and it does not really matter that much: There is not enough time to go to such theoretical and philosophical lengths. While we would advocate simplicity and limiting of scope wherever possible, we cannot support such a brush-off. The devil is in the detail, and at some point, if you have not modeled your data well enough to be self-explanatory, someone is sure to reuse it incorrectly. The Semantic Web helps to guard against such misuse through misunderstanding.

2.4.4 DATA PUBLISHING AND DISCOVERY

The main benefit of Linked Data is that it provides a standardized way of publishing data on the Web that makes it easier to find. By instituting a recognized format in which to publish the data, semantic search engines can follow links from one dataset to another, and hence discover new information, in the same way that documents are linked together on the World Wide Web. The difference is simply that instead of the links being between pages describing things, the links are actually between the things themselves. While search engines built explicitly on semantic technology, such as Hakia,[9] Sindice,[10] and SWSE,[11] can use the Linked Data to find results of your search in context, publishing data as RDF can also push your Web site up the Google rankings. This kind of search engine optimization can be hugely important for organizations trying to gain brand recognition or sell their data. One example of a site successfully rising up the Google rankings is the BBC's natural history site, which is now embedded with semantic tags. Try searching on Google for an animal, vegetable, or other natural history topic—say the sunflower starfish. The BBC Web page is now far more likely to appear in the top few sites returned by Google (at the time of writing, the BBC was the fourth top site for the sunflower starfish).

It can make good business sense to publish your organization's information as Linked Data. For example, in 2009 Best Buy launched a Semantic Product Web beta, using the GoodRelations[12] e-commerce ontology with RDFa (Resource Description Framework-in-attributes) microformat tags embedded in the pages and saw a 30% increase in traffic to the Best Buy pages.[13] A number of potential business models for Linked Data have been discussed by Scott Brinker,[14] from subscription-based services, through advertising, to traffic generation and increasing brand awareness. We discuss these ideas and others in Chapter 8 when we look at how to generate a return on the investment in curating and publishing Linked Data.

Linked Data also offers the benefit of assisting data providers to build services based on their content, which may be easier to monetize than trying to sell their data directly—particularly as Web users expect more and more data to be offered for "free," and in the GI field, with Google Maps, a basic level of GI data already is widely used without direct payment. If your GI data is easily available in RDF format, it is also easier for other people to build applications or services based on your data, particularly if they are using technologies outside your core competency, such as mobile apps, which can generate an indirect revenue stream for you. As always, getting in there early pays dividends. By being early to market in your sector, your Linked Data set can become a hub to which many other datasets link, increasing your discovery rate.

Another benefit of Linked Data is that of trust and authority. Since every data item is identified with a Uniform Resource Identifier (URI), the source of the data is immediately clear. Although, as described further in this chapter, the Semantic Web technology stack layer of Trust has yet to be fully developed, the ability to link back to an original source is a first step in this direction.

Last but not least, Linked Data allows search results to be more dynamic. As new data sources are published on the Web of Data, answers returned by semantic queries become more *complete*, as they can include elements of the new data sources, rather than merely that a new Web page pushes an older one further down the search results.

2.5 HOW IT WORKS

The Semantic Web is based on a number of related technologies, as shown in Figure 2.2. This section is intended just to introduce the names and roles of the various technologies so that the reader will survive the acronym overload that is to come. The most significant technologies are discussed in more detail in further chapters, but the main ideas are discussed next.

The bottom layer in the stack is made up of the Unicode character set for universal encoding of text on the Web, along with identifiers that uniquely identify each data item on the Web. Identification of things on the Web (known as "resources") is achieved by the use of URIs. The well-known URL (Uniform Resource Locator) or "Web address"

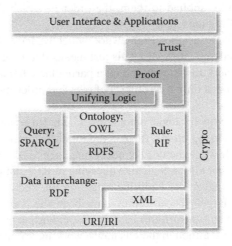

FIGURE 2.2 Semantic Web technology stack. IRI (Internationalized Resource Identifier) = a superset of the URIs, which can include characters from the Universal Character Set (that is, including characters from languages other than English). (From http://www.w3.org/2007/03/layerCake.png Copyright © 2007 World Wide Web Consortium, http://www.w3.org/ [Massachusetts Institute of Technology, http://www.csail.mit.edu/; European Research Consortium for Informatics and Mathematics, http://www.ercim.org/; Keio University, http://www.keio.ac.jp/]. All rights reserved.)

is a kind of URI, but a URI is a more general term that encompasses any item on the Web. The other sort of URI is a URN, or Uniform Resource Name, which gives the name of the resource without explaining how to locate or access it. For example, http://www.ordnancesurvey.co.uk/oswebsite/partnerships/research/ is a URL, while os:BuildingsAndPlaces/Hospital is a URN, and http://data.ordnancesurvey.co.uk/doc/50kGazetteer/218013 is a URI for the data resource Southampton in Ordnance Survey's RDF data.

The next layer in the stack of Figure 2.2 is XML, the eXtensible Markup Language, which is a W3C standard for marking up or tagging data or documents. The following is a very simple XML document:

```
<?xml version = "1.0" encoding = "UTF-8" ?>
<data>
    <sentence lang = "en"> Here's a sentence. </sentence>
</data>
```

The tags are all enclosed in angle brackets <>, with a backslash to indicate the end of that particular markup. lang is an attribute denoting which language is used, in this case, English. The tags that are allowed in a particular XML file can be specified in a Document Type Definition (DTD) or XML Schema Document (XSD). This is a listing of which tag names, structures, and attributes are valid for a particular document. In the example, the DTD might state that <sentence> can only be used within the <data> tag. A DTD is an example of a *schema*, also known as a grammar. A schema constrains the set of tags that can be used in the document, which attributes can be applied to them, the order in which they appear, and the allowed hierarchy of the tags.

RDF, the Resource Description Framework, which we have already mentioned, is often said to be "serialized" in XML. This just means that RDF data uses the XML tag structure for its markup and is based on a particular schema—in this case RDF Schema (RDFS)—so that only a certain set of tags and ordering is permissible in an RDF file.

At the next layer up sits the logic: ontologies, which can be described using OWL, and rules, for which, among others, there is the RuleML family of languages (Boley et al., 2011). Alongside these there is the query language SPARQL (Prud'hommeaux and Seaborne, 2008) that allows SQL-like querying of RDF data (although not OWL instances). The higher layers, of Unifying Logic, Proof, and Trust, are notably free of acronyms; this is largely because they have been the least researched, and the W3C (World Wide Web Consortium) has yet to standardize any languages or tools to address the problems, although there is now a working group in the area of provenance. It is the Trust layer that relates to provenance: explanations of why a particular result has been returned as the answer to a query, where it has come from, and how reliable it might be. There have been early discussions about "semantic spam," where incorrect markup is maliciously added to data, or erroneous links made, resulting in incorrect answers to queries or misdirection of users' searches. Chapter 8, which discusses the publishing of information as Linked Data, explains in more detail some of the strategies that spammers could take and what to watch for.

2.6   RECENT TRENDS IN THE FIELD

Recent trends in Semantic Web technologies have taken several different avenues. From the "Web 1.0" direction, there has been incremental change toward embedding structured information into Web pages, using microformats. Microformats reuse standard XHTML tags to express metadata, which can provide a low barrier to entry for organizations that are put off by the complexities of description logics and knowledge modeling in heavier-weight Semantic Web technology. For those who are reluctant to invest in a completely different way of managing their business intelligence or content, the microformat option offers a first, lower-risk step toward greater semantic understanding of their Web content, based on the "pave-the-cowpaths" principle.

There are several microformats for encoding geographical information, for example, to mark up WGS84 geographic coordinates (latitude and longitude),[15] or encoding information about waypoints,[16] which can be embedded in HTML, XHTML, or syndication formats like Atom or RSS. Although it lowers barriers to entry, there are severe limitations to taking the microformat approach, which can be seen in the kind of discussions that are taking place on wikis developing the geo extensions to currently available microformats.[17] The contributors are not necessarily GI domain experts, and they find themselves falling into a trap of scope creep. It is often not clear what the specific purpose of these additional markup tags are, and because they are limited to HTML, which is a document format not a data format, there can be no explanation of what each markup tag means or how it should be used. In other words, more semantics are needed and a more disciplined approach to knowledge modeling.

The RDF community has answered this plea to lower barriers to entry by the development of RDFa. RDFa is a W3C recommendation that enables the embedding of RDF data within XHTML documents based on the microformat approach of coopting the XHTML attributes. This means that RDF-compliant software agents can extract this data from the XHTML attributes and understand the semantics in the data. One such example of this type of search agent was Yahoo! Search Monkey.[18] Search Monkey used the RDFa metadata to enrich search result display and provided a developer tool to extract data and build apps to display the Web site owner's own custom applications. Although this closed in October 2010, the reasoning behind the change was that SearchMonkey required developers to build a lightweight application or service using the Search Monkey API, whereas Yahoo!'s new strategy is to encourage Web developers to embed RDFa data directly in their Web sites, and Yahoo!'s standard search can take advantage of this structured information directly, rather than through an additional application that must be built separately for each dataset. So far from the semantic experiment being regarded as a failure at Yahoo!, this shows how it is moving into the mainstream. Another sign that the need to structure information is being more widely recognized is that schema.org, a collection of markup recognized by the major search engines (Microsoft Bing, Yahoo!, and Google), now includes a subset of RDFa.

A second trend in Semantic Web technology is an expansion of the infrastructure required to support the Web of Data: development of the RDF databases (known as "triple stores" as they are designed to hold triples). Several commercial ventures, from well-known players in library science (Talis[19]) and database management (Oracle[20]), as well as other software companies (Open Link Software[21]) have been active in this space, and efficient management of large-scale datasets remains integral to the development of the Web of Data. We discuss in more detail how these triple stores work in Chapter 7.

Another trend of more immediate interest to the GI professional is the opening up of government data, including geographic data, in both the traditional sense of using Web APIs and as linked RDF data. Examples of this are the United Kingdom's data.gov.uk program, which has to date published over 5,400 datasets, from all central government departments and a number of other public sector bodies and local authorities, and the data.gov site in the United States, which has made over 6.4 billion RDF triples of U.S. government data available for exploitation. Further sources, like the Europe-wide publicdata.eu and the Swedish opengov.se, have also been published without direct government support. The aim is to foster transparency in government by making the data more easily available and searchable, to allow cross-referencing or "mashing up" of the data and semantic browsing or SPARQL querying across datasets.

Recent developments in geosemantics have included suggestions to incorporate spatial logics such as Region Connection Calculus (Randell, Cui, and Cohn, 1992) for qualitative spatial representation and reasoning as a semantic technology (Stocker and Sirin, 2009), as well as a proposal for GeoSPARQL (Perry and Herring, 2011). GeoSPARQL is a spatial extension to the SPARQL query language and allows simple spatial queries to be formed using point, line, and polygon data types. It is discussed further in Chapter 8.

As mentioned, we predict that the topic of data provenance will become increasingly important as the number of RDF datasets published on the Web increases. As witnessed by the current Web 1.0, it can be difficult to confirm the veracity or accuracy of anything on the Web, and in the semantic sphere, this is even more the case as context also plays a significant role: The data might have been correct or useful for the purpose for which it was originally captured, but it can be harder to answer the question of whether it is still valid when reused in a slightly different context. There are technical difficulties in how to add information about provenance or context to an RDF triple as metadata, which are discussed in Chapters 5 and 8, but resolving these issues remains an important step in the development of the Web of Data.

While it has frequently been described as orthogonal to the Semantic Web, the final trend that we wish to mention here is the Social Web, or Web 2.0. Although often depicted as being about social networks, the Web 2.0 trend is really about the Web going back to its roots as the "Read-Write Web," which has facilitated a huge increase in user-generated content. There are several ways in which this user-generated content can be used on the Semantic Web. The most straightforward way is for the information on user-generated Web pages like Wikipedia to be scrapped and published as RDF (the RDF version of Wikipedia is called DBPedia[22] and contains both GI and non-GI-related data). Other options are to exploit the semistructured information available from user-authored tags such as the image and video tags on Flickr[23] and such as the research carried out at France Telecom using natural language-processing tools to automatically convert a set of tags into an RDF description (Maala, Delteil, and Azough, 2007). These loose hierarchies generated by users are often called "folksonomies" and are effectively a method of crowd-sourcing ontologies—with all the accompanying issues of accuracy, disagreement, bias, and currency that this entails. Some sites like Foursquare[24] ask users to input data within a fixed hierarchy, which is more easily converted to RDF. Users are willing to do this as the data they add represents an integral part of the application. The exploitation and linking of this type of user-generated, semantically enriched data is a field ripe for development at the intersection between the Semantic Web and the Social Web.

2.7 SUMMING UP AND SIGNPOSTS TO THE NEXT CHAPTER

This chapter has covered a lot of ground to lay the foundations for our coming discussion of the geographical Semantic Web. We explained the relationship between the traditional Web of documents and Web of Knowledge that the Semantic Web represents. The concept of Linked Data has been explained, and we positioned it within the context of other Semantic Web technologies and the history of their development. Discussion was devoted to the benefits of the Semantic Web for both organizations and individuals, including potential business models for exploiting Linked Data. By describing the basic technology stack, from identifiers and character encoding at the bottom, through to Trust and the user interface at the top, we have provided grounding in the terms that the reader will encounter in subsequent chapters.

Now, we move on to discuss GI as it is today, summarizing its successes and struggles. We also touch on how crowd-sourced geographic data challenges the GI professional and how the two can fit together in the future.

NOTES

1. This axiom is written in the controlled natural language Rabbit; a brief description of Rabbit is given in Appendix B.
2. The methodology for authoring ontologies discussed in Chapter 10 includes a step to specify scope, specifically to reduce the likelihood of this problem.
3. http://www.geneontology.org/.
4. GeoFeatures ontology, http://www.mindswap.org/2003/owl/geo/geoFeatures.owl
5. http://ontologydesignpatterns.org
6. http://www.ordnancesurvey.co.uk/ontology/spatialrelations.owl
7. http://www.geonames.org/.
8. Linking Open Data cloud diagram, by Richard Cyganiak and Anja Jentzsch. http://lod-cloud.net/.
9. http://www.hakia.com/.
10. http://sindice.com/.
12. http://swse.org/.
13. http://purl.org/goodrelations/.
14. As reported in http://www.chiefmartec.com/2009/12/best-buy-jump-starts-data-web-marketing.html
15. http://www.chiefmartec.com/2010/01/7-business-models-for-linked-data.html
16. http://microformats.org/wiki/geo
17. http://microformats.org/wiki/geo-waypoint-examples

18. http://developer.yahoo.com/blogs/ydn/posts/2008/09/searchmonkey_support_for_rdfa_
    enabled/.
19. http://www.talis.com
20. http://www.oracle.com
21. http://www.openlinksw.com/.
22. http://dbpedia.org/sparql
23. http://www.flickr.com
24. https://foursquare.com/.

3 Geographic Information

3.1 INTRODUCTION

This chapter introduces the concept of Geographic Information (GI) and is intended for those not familiar with GI or Geographic Information Systems (GIS). It describes the main types of GI and how they are used, which is an essential element in the understanding of GI. We then provide a short history of GI because it is not only necessary to understand the forms and uses that GI takes but also to understand how it has developed. This history shapes both how GI is used today and how it is viewed. For those who do not consider themselves to be GI experts, this chapter provides a necessary background for the rest of the book. For those who are knowledgeable about GI, it may still introduce them to a broader view of GI than they may be yet familiar with and will open up the opportunity to think differently about this topic.

3.2 WHAT IS GEOGRAPHIC INFORMATION?

Geographic Information is quite literally all around us; it is a part of our everyday lives. We use GI when we post a letter to a friend or when we navigate to work or watch the news about a foreign conflict or a local planning dispute. GI is used to determine who we get to vote for, who provides our local services, and where a vendor chooses to place a new shop. It is involved in understanding who we are; it helps to set our self-identity and defines the sides in sporting events and wars. GI is also used when we monitor our environment and understand the natural world—enabling us to realize the damage we do and offers hope that we can correct some of our mistakes. It is used to track the progress of disease and to spot causal links between industrial pollution and ill health. GI runs through our history and is one of the boundaries that determine that very history. GI is so much a part of our lives that we do not notice it. Most of us take it for granted. But, what exactly is it? In one sense, this is a very easy question to answer: It is any information that references the geography of the world. In another sense it is a quite difficult question to answer simply because the boundaries of what is and is not GI can be somewhat ambiguous. Most people would agree that the exact position of a ship at sea would be GI; but what about a customer record that contains the customer's contact address as well as other details? Is just the address the GI, is it the address plus the customer's identity, or is it the whole record? And, what about a series of GPS positions that are used for local analysis and that are not directly related back to any visible object on the ground? The answers to these questions are often context dependent. If I am reading a restaurant review, then I might think of the information as GI if I am in the vicinity of the restaurant and am looking for somewhere to eat. However, if my context were different, and I were at home, reading about a restaurant I had previously visited and comparing other people's opinions of the restaurant, it would not fall under the "GI" category. In the former case, it may be treated as GI because it is helping to solve a problem with a geographic component—the *where* in "where and what shall I eat?" In the latter case, there was no *where* to answer. This example highlights that we are most interested in information that can answer questions containing a location component. This book therefore treats GI in its broadest context and leaves it up to the reader to restrict the definition to specific personal circumstances.

GI is often treated in a much narrower sense, as information that has explicit location data associated with it, most often geometry that relates to a particular geographic coordinate system. This is the way that GI is traditionally viewed within what we will term the "GI community"—those (almost exclusively) professionals who specialize in spatial analysis or related areas. However, there is a bigger community of those who may not view GI as the focus of their attention but who nevertheless need to use GI to achieve their aims. The majority of this group have probably not heard of the term *GI*, and their nature and motivations vary so much it would also be difficult to identify them as a single community. This book not only focuses on the GI community but also is relevant to the broad body of people using GI whether they are conscious of it or not.

3.3   THE MANY FORMS OF GI

It would be almost impossible to identify all forms that GI may take if we simply apply the broadest possible definition. However, the vast majority of GI, or at least that essential part of GI that relates to location (whether relative or absolute), can be represented using a number of well-established forms. For the sake of simplicity, we have divided these into three groups: those related to geometry; those related to topology and mereology;[1] and those that are textual. There are other representational forms, such as triangulated irregular networks (TINs) but that are specialized forms of geometry and not the focus for this book (for an introduction to these topics, see Longley et al., 2001).

3.3.1   GEOMETRY

To dispel any sense of anticipation that the reader may develop, the geometric forms of GI are the least interesting from the perspective of Linked Data and the Semantic Web. Nevertheless, including them and giving the reader an understanding of their nature is important to this narrative since it is these forms that have traditionally set GI aside from other data. And, less important does not mean unimportant or irrelevant to the Semantic Web. There are a number of different geometric models that can be used to capture geometric information, the most significant of which are described in the following paragraphs.

3.3.1.1   Raster

In raster form, GI is represented as an array of cells (or pixels), with each pixel representing a value and the position of each pixel corresponding to an area on Earth's surface. GI raster data (rather than raster maps) usually take the form of photography, or some other form of area sensor values, where each pixel represents a particular

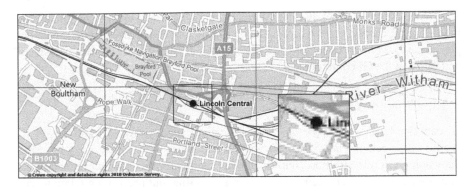

FIGURE 3.1   An example raster map.

hue, brightness, and saturation; or area representations where pixels encode specific aspects of the area, such as land cover with pixel values perhaps indicating trees, scrub, water, and so on; or a height, where the pixels may represent heights. The latter form of raster is also known as a digital elevation model (DEM). Digitized maps are also represented as rasters. Figure 3.1 shows a typical raster map for which the image becomes pixelated if enlarged.

Because the pixels are spatially related, it is possible to perform sophisticated spatial analysis on raster data. Attempting to perform analysis on many raster maps is more problematic as the maps have often been generalized and do not accurately relate to the surface of the earth.

3.3.1.2   Vector

Vector data represents GI as points, lines, and polygons. Although other representations are possible, these three are the fundamental vector forms and the most common. These vector forms are mapped to Earth's surface using coordinates that also have attribute information associated with them to describe the object being represented.

The position of an archaeological find could be represented as a coordinate with associated attribute data identifying the find and perhaps the date of discovery and the estimated age of the object. Real-world objects[2] can be represented by one or more vector objects and forms depending on the complexity and resolution of the model. A building could be represented as a single point, one or more lines, or one or more polygons and indeed any combination of point, line, and polygon. An example of how a simple building may be represented at various levels of detail is shown in Figure 3.2.

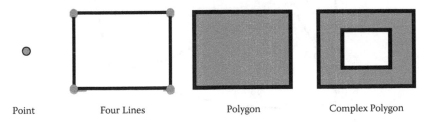

Point   Four Lines   Polygon   Complex Polygon

FIGURE 3.2   Building represented by a point, lines, and polygons.

From a databasing perspective, vector models require special two-dimensional indexes to efficiently retrieve data based on a spatial search. This need for more complex indexes is one of the factors that has made GI appear special, both to "outsiders" and those in the GI community.

3.3.2 Topology and Mereology

GI can also be modeled topologically and mereologically. Topology expresses the spatial relationships that exist between different objects, and mereology describes the whole–part relationships. Neither topology nor mereology provides precise location information; rather, their focus is on representing and preserving relationships. Topology is often used to represent networks such as road or rail systems, where the important aspect is the connectivity between the network components. If we take the example of a road network, we can topologically describe the relationship between one road stretch and another as shown in Figure 3.3. In these topological models, the road network is symbolized as a series of links, representing stretches of roads, and nodes, representing junctions between stretches. We are then able to model the fact that road stretch A links to junction B, which in turn also connects to stretch C.

Topology can also be used to define the relationships between vector objects. For example, if the outline of a building is constructed by separate line features, then these can be topologically related as shown in Figure 3.4.

Mereological relationships can be used to express the fact that a particular building is "part of" a hospital or that a road network "comprises" roads A, B, and C. There are often strong relationships between topological relationships and mereological relationships; if we know that a building is "within" (topology) the grounds of a hospital, it may be reasonable to assume that the building is also "part of" (mereology) the hospital. Figure 3.4 is also an example of not only topology but also mereology. That topological and mereological relationships can often be co-occurring has resulted in the term *mereotopology*. Since mereotopological relationships are commonly

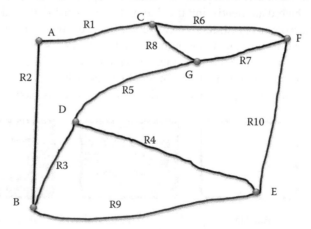

FIGURE 3.3 A simple road network showing the links (R1–R10) and nodes (A–G).

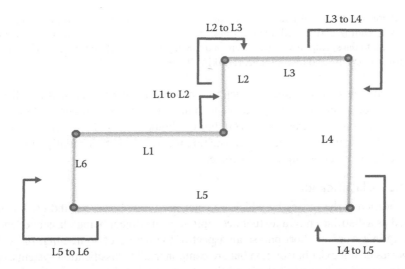

FIGURE 3.4   Building footprint represented as a polygon constructed from multiple lines.

occurring, care must be taken not to confuse topology and mereology. Consider the following: An English county is an administrative unit that wholly contains a number of boroughs or districts, which are also administrative areas. We can certainly say that topologically a district is "contained within" a county. But it is only valid to say that a borough is "part of" a county if there is some formal relationship that makes a borough a functional part of a county. It turns out that this is not the case; essentially, a borough shares an area of land with a county, but each operates independently and has a different administrative role. As we shall see further in this book, the process of describing such objects for the Semantic Web can often tease out such subtleties.

As we will see, both topology and mereology have important roles to play when expressing GI in Linked Data terms.

3.3.3 TEXTUAL REPRESENTATIONS

We are all familiar with textual representations of GI, although we are so familiar with these forms that we tend not to think about them as representations of GI. There are at least four types of such textual representations: description, classification, direction, and address.

3.3.3.1 Description

Textual descriptions are quite simply that: descriptions of the landscape, place, or other geographical feature. Here is a typical example that is simply littered with geographical references:

> As Barry, Jane and I walked up the track across the heath to our old campsite, we passed Gentian Valley, and, reassuringly, the deep blue marsh gentians were still there, in flower, half hidden in the heather. But gorse had taken over the campsite hollow,

so there was no clue it had ever been there. We crossed the wooden bridge over the railway cutting and turned downhill past the Scots pines on Black Down, following the path to the spring. It was still there and flowing well enough, but fenced off by the railway company and inaccessible. (Deakin, 2008)

Descriptions can be extremely rich in terms of information content but difficult to access from a machine perspective due to the use of natural language. As a result, such descriptions are rarely seen as GI, but we should not ignore this type of content as it can be either text mined for information that can then be formally represented or stored as is for consumption by people.

3.3.3.2 Classification

Classification is the means by which we categorize the objects around us. It can be viewed as a distillation of a textual description of the object being classified, encapsulating in a word or short phrase an aspect of the nature of something. Individual categories do not exist in isolation but are components of classification systems, typically taxonomies: hierarchical assemblies of categories adhering to some system of classification, or simple lists of categories, often known as "controlled vocabularies." Traditionally, classification schemes were the remit of the professional, carefully constructed and applied by the professional for the professional. Where the public was exposed to them, it was based on viewing the classifications that had been applied. Web 2.0 has changed that to an extent, opening up the classification of the world to the hoi polloi (although in truth there has always existed a tradition of classification systems being constructed by the expert amateur). This has resulted in more informal classifications arising, often termed *folksonomies*. A folksonomy is an emergent system of classification that arises when individuals create categories (or tags as they are often known by the contributors). The folksonomy is the resultant collection of tags, which may be hierarchically structured (but more often not), representing some form of consensus. Tagging is also used even more informally to classify anything from photographs like in Flickr, to destinations, as in Foursquare (although here there is an attempt to develop a hierarchy). In some cases, there is a high degree of agreement regarding the meaning of the classes or tags; in other cases, the systems are quite ambiguous, with different people using different tags to mean the same thing or the same tags to mean different things. Professional systems and some folksonomies represent single worldviews that throw up limitations of expression, although they are generally less ambiguous. Other folksonomies are effectively a mosaic representing the views of individual contributors and can therefore be very ambiguous. In either case (professional and amateur), ambiguity can occur and can only be resolved if the semantics of the individual classes are well defined and the classification system described. This is often not the case, or a classification system has only been incompletely applied, leading to the possibility of misinterpretation.

Another form of classification system, introduced in Chapter 2, is ontology. Ontologies can help to overcome some of the problems of ambiguity by more formally defining the categories and their relationships to other categories. Ontologies form an important component of the Semantic Web, and this book therefore deals with them in great detail.

TABLE 3.1
Human and Machine-Generated Directions

Directions by a person	Directions from Google Maps
Turn Right onto Alford High Street drive ahead and at the crossroads turn right onto Graystone road continue along past the hotel on the left until you reach the junction. Turn right onto Main Street and continue out of the village heading toward Aberdeen on the A944. After about 15 miles you will pass the Loch of Skene on your right, and as you approach Westhill keep going straight on through Westhill passing straight through 2 roundabouts. The next roundabout has Tesco [a shop] on the right but go straight, leaving Westhill. 1/2 mile comes to another roundabout and turn left onto the dual carriageway heading toward Kingswells. This next roundabout has traffic lights but go straight ahead 2nd exit and keep in the right-hand lane. After 1/2 mile take the second exit at the junction heading toward Hazlehead with the Crematorium on the right.	1. Head **southwest**, about 1 min, go 0.5 mi 2. Turn **left**, go 0.1 mi 3. Turn **left** toward **A980**, about 2 min, go 1.4 mi 4. Turn **right** at **A980**, about 6 min, go 3.7 mi 5. Turn **right** to stay on **A980**, go 190 ft 6. Take the 1st **left** to stay on **A980**, about 1 min, go 0.6 mi 7. Slight **right** toward **A93**, about 9 min, go 4.8 mi 8. Turn **left** at **A93**, about 6 min, go 3.4 mi 9. Turn **right** at **B993**, go 0.1 mi 10. Turn **left** toward **B976**, about 6 min, go 3.9 mi 11. Turn **right** at **B976**, about 1 min, go 305 ft 12. Take the 1st **left** toward **B974**, about 4 min, go 2.3 mi 13. Slight **right** at **B974**, about 16 min, go 10.1 mi 14. Turn **left** to stay on **B974**, about 2 min, go 1.3 mi 15. At the roundabout, take the **2nd** exit onto **Main St/B966**, go 322 ft 16. Turn left at Burnside Rd/B974 17. Continue to follow B974, about 8 min, go 4.6 mi 18. ….

3.3.3.3 Direction

Directions can be expressed as free text (which can in fact be description), or they can be formal, often machine generated from topological relations. In essence, directions are just a textual way to express certain topological relationships with the aim of connecting a start point with an end point. However, one thing that is apparent is the difference between those directions generated by a computer and those generated by people (Table 3.1).

Both sets refer to the same journey. They describe slightly different routes, but the main difference is the style. The directions as given by the person are rich in references to landmarks, and where distance or time are mentioned, they are often approximate: "after *about* 15 miles." By contrast, Google's directions are, well, very machine like, precise in distance and only approximate about time. The machine instructions also lack any reference to landmarks;[3] the machine does not know what landmarks are, and even if it did, it probably would not have any way of knowing those relevant to this journey.

3.3.3.4 Address

Addresses can be thought of as standardized descriptions of a place such that the addressed object can be located in the landscape. Most countries have some

TABLE 3.2
Typical Address Structure

[Premise]	Flat 5		
Building	10	The Hampshire Bowman	12201
Road	Drum Lane	Dundridge Lane	Sunrise Valley Drive
[Locality]	Spice Island		
Settlement	Portsmouth	Dundridge	Reston
[Region]	Hampshire	Hampshire	Virginia

Note: Square brackets indicate optional fields.

standardized form of addressing, often including a postcode or zip code. Putting aside the postcode or zip code for one moment, most addresses have a structure of the basic form shown in Table 3.2, although there are plenty of variants. A more significant variant can be found in Japanese addresses; buildings are numbered by order of build, not position in a street, and located to city blocks not streets, the blocks being named and the streets unnamed.

However, addresses can be more complex and often do not follow any rigid pattern. The address

High Birch Cottage
High Birch Farm
High Birch Lane
Wheeley Heath
Wheeley
Nr Great Bentley
Essex

contains a reference to both a building (High Birch Cottage) and the farm it belongs to and then includes a locality (Wheeley Heath) as well as a settlement (Wheeley) and then another settlement (Great Bentley) as well as the relationship "Near" (Nr).

Addresses were developed to identify man-made objects, typically buildings, and are therefore less good at identifying natural features. Even many man-made things, such as structures (fixed cranes, pylons, bridges, etc.), do not have addresses. They are also most often connected with the delivery of post, and postal addresses can vary from what might be termed a geographic address (one that is based solely on geographic-topologic relationships). Compare a geographic and postal address for the same location in England:

Geographic address:	Postal address:
5 Green Lane	5 Green Lane
Warsash	Warsash
Hampshire	SOUTHAMPTON
	SO31 X10

In the postal address, the county reference (Hampshire) has been replaced by a post town (Southampton). However Warsash is not within the boundaries of Southampton, but from a postal perspective, the post for Warsash is directed to the main regional sorting office in Southampton. Postal addresses do not always differ from geographical addresses, but one needs to be aware that postal geography can be quite different from what one might term a natural topographical geography. Postal addresses also frequently have a postcode or zip code, again intended to aid the postal service in delivering the mail by enabling addresses to become more easily machine readable. A number of geographically close addresses will all share one postcode. These codes can be used to create their own geographies as they provide easily referenceable labels that can be used to associate information about that postcode area. For example, someone could associate health statistics to postcodes, enabling the health of an area to be mapped without identifying individual addresses and thus protecting personal information. However, it needs to be remembered that postcodes can change, and that they do not really identify areas, just a collection of individual postal delivery points. The area associated with a postcode or zip code can also vary enormously and will be much smaller in cities than rural areas.

3.4 REPRESENTATIONS AND USES OF GI

3.4.1 MAPS

A map, a two-dimensional representation of the landscape, is the most obvious manner in which we think about GI being represented. The most common form of maps is generalized maps that do not show an area precisely but allow clarity of information visualization to take precedence over exact positional accuracy.

In generalized maps, things such as roads are often enlarged (broadened) to emphasize them, unimportant buildings may be removed to reduce clutter, and important buildings may be represented as symbols to highlight them. Another type of map, less well known to the general public, is the planametric map; precision is king, and all things are shown in their correct proportion and relationship to other map objects. Such maps are typically used by builders and utility companies, for land registration and other applications where the exact layout is required. Topologic maps are also another popular representation. Such maps show the relative relationships of the mapped objects but are imprecise in locational terms. The most famous example of a topological map is the London Underground map. It has generated many maps that have adopted its style, such as the Milky Way Transit Authority map shown in Figure 3.5.

Maps are used for a number of distinct purposes, the most common being

- Navigation
- As a backdrop to overlay other information
- As a way to visualize aspects of a landscape

Maps are therefore primarily for information visualization, at least in the traditional sense, and therefore are less relevant to Linked Data and the Semantic Web,

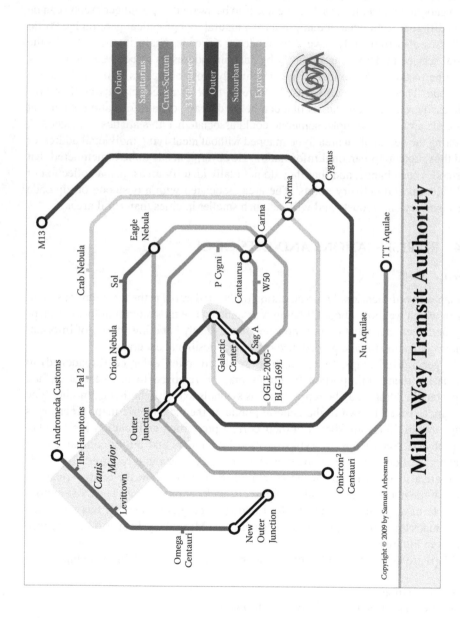

FIGURE 3.5 Milky Way Transit Authority map: a topologic map. (Image courtesy of Samuel Arbesman.)

although these technologies can be used to represent the underlying data used to construct the map.

3.4.2 GAZETTEERS

A gazetteer is an index of places (or other spatial objects) that in its most basic form lists the object against its position. It may also provide some information about the object. For example, the online gazetteer for Scotland (http://www.scottish-places.info/) describes itself as "a vast geographical encyclopaedia, featuring details of towns, villages, bens and glens from the Scottish Borders to the Northern Isles." Each entry includes a description of the place and its position; an example is given in Figures 3.6 and 3.7. It may also have associated photographs and information that relates it to other places, for example, those that border it.

For many years, gazetteers have been seen as somewhat of a backwater in GI: useful indexes at the back of atlases or maps but not something to cause excitement. More recently, increasing interest has been shown in gazetteers, and they are undergoing a period of renaissance. Gazetteers in the digital age can be more than just indexes to maps but indexes to resources, including both data and service resources. As a consequence, they have an increasingly important role to play in the age of digital GI and Linked Data.

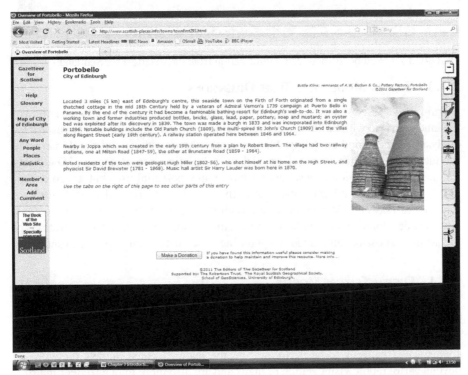

FIGURE 3.6   Portobello: an example of an entry from the gazetteer for Scotland.

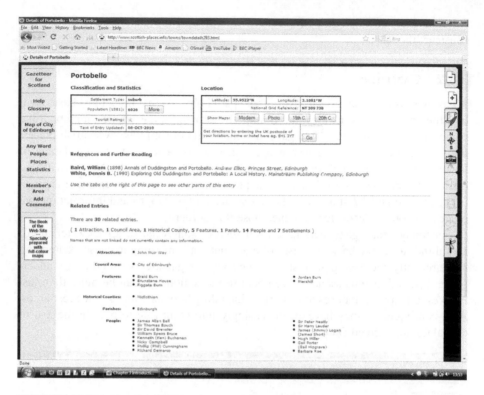

FIGURE 3.7   Portobello: more detail information from the gazetteer for Scotland.

3.4.3   TERRAIN MODELS AND THREE DIMENSIONS

Terrain models are used to represent the height of a landscape, typically the height of the land's surface with buildings and vegetation removed. Such models are useful for analysis, such as flood modeling. More recently, there has been a general interest in producing three-dimensional (3D) models with buildings represented in 3D to varying degrees of precision. These models have many potential uses but are so expensive to produce that, at least at the time of this writing, economically viable uses are few and far between. At present, where such models exist, they are mostly to show how new developments might fit into an existing landscape or are demonstrations of the art of the possible and to aid the planning of security at large public events.

3.4.4   DIGITAL FEATURE MODELS

A digital feature model is a representation of the geography of an area, specifically discrete landscape features and their relationships to each other, typically constructed using vector and network data. The data is stored in a database, allowing sophisticated analysis to be performed on the features. Such models therefore go beyond cartographic visualization of maps and are also more than gazetteer indexes. Feature models enable a vast range of possible uses for GI, including environmental analysis, routing, health analysis, insurance risk calculations, placement of retail parks, and so on. The list is almost limitless. The important factors of any digital feature model are identity, a means to uniquely identify the feature; location, a means to locate the feature in the landscape (normally achieved via vector geometry associated to a coordinate system); classification, which specifies the nature of the thing being represented; and attributes, which may include relationships to other features. If one is precise, then the locational information and classification are also attributes, although they are of special interest.

Digital feature models lend themselves to representation as Linked Data within the Semantic Web.

3.5   A BRIEF HISTORY OF GEOGRAPHIC INFORMATION

To understand the current state and usage of GI, it can help to understand its history as past decisions have an impact on the shape of today. GI in its widest sense has a history that goes back way before the age of computers to the waggle dance: where bees have communicated the location of pollen through their movement. But, this book need only concern itself with the use of GI within computer systems and the Web.

Giving a brief history of GI usually means giving a history of the professional aspect of GI, which, to respect tradition, we also do. But, there is also a more informal use of GI, whose own history is also worthy of telling and indeed becomes more relevant as time progresses. The reader should therefore excuse the fact that two histories are presented with only a small degree of interaction between them. Part of the future for GI will be to bring these two stories together.

3.5.1   A Traditional Story: GI and GIS

The standard history of GI is not really specifically about GI, but about GIS—Geographic Information Systems. GIS explicitly recognize that GI needs special treatment and provide technologies for performing specialist analysis. It is also easier to tell the story of GIS: It is easy to search the Web for a definition of GIS, but searching for a definition of GI is much less fruitful; at the time of writing, Wikipedia, that global penny bazaar of knowledge, had no entry for GI. It may even be that the term GI did not exist independently before GIS.

3.5.1.1   Geographic Information Systems

The first accepted reference to GIS is the Canada GIS built in 1963, as mentioned by Longley et al. (2001). The important aspect of this system was the explicit recognition that computers, which in 1963 were very novel, could be used to perform spatial analysis. The system was used to store maps that resulted from a survey of the potential use of land for various purposes, such as agriculture or forestry. The system developed such that it acquired the capability not only to hold the maps but also to perform analysis on them through a series of fixed reports. This may seem primitive by today's standards but was revolutionary for the time. The late 1960s saw the U.S. Bureau of Census also developing a GIS to create a street gazetteer to support the collation of census records. The U.S. and Canadian systems were not related but shared certain common aspects in the manner in which they handled GI. Over time, the need to address GI in a common way became more clearly recognized, leading to the development of commercial off-the-shelf general-purpose GIS in the 1970s.

Larger national mapping agencies, such as the U.S. Geological Survey (USGS), Ordnance Survey of Great Britain, and Institut Geographique National (IGN) France also recognized that computer systems could be used to help the map production process. The systems they produced were not strictly GIS since they were designed to manage and edit map data for cartographic purposes. Nevertheless, there were obvious similarities between systems designed to perform analysis and those with more cartographic intentions. What was common to all these systems was the ability to manage, manipulate, and index geometry and to a lesser extent topology.

By the 1980s, commercial GIS were emerging, with companies such as ESRI (Environmental Systems Research Institute, Inc.), which launched ARC/INFO in 1982, and competitors such as Integraph, CARIS, and ERDAS also developing commercial GIS software in the early 1980s. In 1982, there also was the emergence of GRASS (Geographic Resources Analysis Support System) GIS, a public domain GIS. All these GIS were hosted either on specialist hardware or on UNIX-based workstations such as SUN and Apollo or other minicomputers, such as the DEC (Digital Equipment Corporation) VAX ranges. In all cases, the solutions were extremely expensive, limiting their market penetration and resulting in the development of GIS specialist groups within organizations.

GIS were also only used in specialist areas. At their simplest, they were used to digitize mapping, sometimes performing transformations between different map projection systems. Other uses were GI visualization, for example, to show the distribution of forestry in an area, perhaps differentiating between different types of forestry; and spatial calculation, for example, to estimate area coverage by forest type. More sophisticated still, GIS enabled organizations to perform analysis such as determining objects within a certain distance of another object or to answer questions like "How many sightings of a particular species have occurred within 500 m of a river?" To provide such capabilities, GIS needed to support a number of core functions, all of which were present in these early systems. These core functions included

- Representing geographic objects in terms of simple geometries: points, lines, and polygons
- Representing GI as a raster (an array of data points representing discrete values)
- Associating other data to these geometric objects
- Describing geometry in terms of one of a number of coordinate systems
- Indexing the geometry to allow selection based on position
- Performing basic spatial analysis, such as identifying objects within another object, objects that are touching or overlapping, and objects within a given distance of other objects
- Visualizing GI, typically as a map display or tabulation

By this time, the idea of GIS layers had also been established—essentially a means to organize data in a number of layers that could be overlaid on top of each other to visualize the information. The layer concept also allowed the division of information into categories, as a layer's content was determined by geometry type and type of object. So, a GIS would have a layer of line objects that might represent roads; another of polygon objects that could represent forestry, lakes, and urban areas; and perhaps another polygon or raster layer representing different soil types. Analysis could be performed using the GIS to answer questions such as, "Which areas of forestry contain a particular soil type and are within 10 miles of a lake or river?"

All of these early GIS were also entirely proprietary in terms of their implementation. Each had its own data formats and structures, each being mutually incompatible with all the others. This was a time when it was also not unknown for organizations to develop their own entirely specialist GIS for their own particular purpose, so a local authority might develop a GIS to manage a land terrier, a utility company might have developed its own solution for asset management, and so on. The 1980s and indeed the early 1990s were not noted for interoperability.

By the early 1990s and with the introduction of the IBM PC (personal computer), the GIS was becoming more affordable, although it still tended to reside largely within GIS departments. During this period, the functionality rose, and the GIS transitioned from workstations to PCs. The market penetration of GIS also widened as vendors developed more lightweight GIS. ESRI, with its industrial-strength GIS ARC/INFO, launched a product called ArcView, designed for the PC market, which was compatible with ARC/INFO but had comparatively limited functionality and was primarily aimed at visualizing GI. The early to mid-1990s also saw rising interest in GIS from the main database vendors such as ORACLE, IBM, Informix, and Ingress. They all developed spatial indexes and extensions to the SQL (Structure Query Language) to enable spatial queries and simple GIS functions to be performed. The establishment of GI capability within large-scale database systems not only increased the volume of data that could be held but also meant that GI started to be managed in the same way as any other corporate data.

The early 1990s also saw the development of new GIS based around the object-oriented (OO) design and programming paradigm, two such examples being the Smallworld GIS and GOTHIC from Laserscan. These attempted to break free from the idea of GIS layers where the main driver was the type of geometry. This meant that they tried to more closely model the nature of a real-world object. In these models, topology was as important as geometry. However, such systems have as yet failed to gain widespread appeal, in part because of the investment companies have made in the more traditional GIS models. OO GIS have been sidelined into niche markets; for example, Smallworld specializes in networks, largely for utility companies, and Gothic serves digital map production systems.

As the capabilities of GIS advanced, so did take-up increase. Today, GIS is firmly established in specialist GIS sections of central and local government and in many commercial companies. It is used in a wide range of applications, from the traditional uses such as preparing maps, to a wide range of applications covering everything from mineral resource discovery to assessing insurance premiums and risk.

Beyond GIS, the use of GI also has grown in industry and government as a means to integrate different data using addresses (most usually postal addresses). The reasons for such integration are many and varied and include integrating data following company or government departmental mergers; fraud detection and credit checking; marketing and general development of a demographic picture; public health awareness;

and so on. The growth of this market is an indicator of the importance that geography plays as a common factor between different datasets. When it has been established that data collected by one organization about a particular place refers to the same place as the data collected by another organization, the data can be combined.

3.5.1.2   Standards Develop

At around this time, there was another significant advance: industry standards started to be developed. Even by the mid-1990s most GIS could at most interoperate on the level of importing or exporting the data formats of other vendors. Standards for GI began to develop in part as a response to the lack of interoperability between GIS and in part as a need to ensure existing standards such as SQL did not fragment. The Open Geospatial Consortium (OGC) has been the main driving force behind standards development. Formed in 1994, OGC has grown to be an industry body with a membership of several hundred organizations and universities with an interest in GI. The Simple Feature[4] Model, Geography Markup Language (GML) (International Organization for Standardization [ISO] 19136:2007), Web Map Server (WMS) standard (ISO 19128:2005), and the Web Feature Server (WFS) standard (ISO 19142:2010) are probably the four most significant standards to have arisen from OGC. All of these standards were emergent in the early 2000s and were influenced by the GIS Layer model, the OO paradigm, and the need to transport GI across the Web. However, they were largely written by and for the GI community—those interested in GIS analysis or producing digital mapping of some form or other in a professional context. Hence, they are an inwardly looking set of standards to be used within a community and are not well known beyond it. Due to their origin and focus, the standards are also geometry-centric by design. As a result, their uptake has been largely restricted to the GI community. In comparison, the Keyhole Markup Language (KML), which serves a similar purpose to GML, was created outside the GI community[5] and is very popular with those wishing to exchange mapping data. KML has also become an OGC standard, but only after KML had already become very well established among nonprofessional users as well as many professionals outside the GI community.

3.5.1.3   The Web

In 1989, Tim Berners-Lee, a computer scientist working at CERN (European Organization for Nuclear Research), proposed what would become the World Wide Web (Berners-Lee, 1989), and in 1990 along with Robert Cailliau proposed Hypertext to link documents on the Web (Berners-Lee and Cailliau, 1990). From that point, the dramatic rise of the Web has been well documented. It was not long before mapping began to emerge on the fledgling Web. Possibly the best-known early example was the Map Viewer application built by Steve Putz at Xerox PARC (Palo Alto Research Center) in 1993 (Figure 3.8) (Longley et al., 2001). Looking very crude by modern standards, it nonetheless had all the features that we consider to be essential to any modern map viewer, including pan and zoom functionality.

People also began to incorporate simple maps into Web-based applications; for example, trade directories could show a map of where a particular service provider was located, local authorities showed development plans, and environmental bodies were able to report on flood risk areas and pollution. GIS companies were also quick

---

Example Showing the Default World Map View

The links on this page connect directly to the Map Viewer at Xerox PARC.

Map Viewer: world 0.00N 0.00E (1.0X)

Select a point on the map to zoom in (by 2), or select an option below. Please read About the Map Viewer, FAQ and Details.

Options:

- Zoom In: [2], [5], [10], [25]; Zoom Out: [1/2], [1/5], [1/10], [1/25]
- Features: Default, All; +borders, +rivers
- Display: color; Projection: elliptical, rectangular, sinusoidal; Narrow, Square
- Change Database to USA only (more detail)
- Hide Map Image, No Zoom on Select, Reset All Options

Options can also be typed in as search keywords (e.g. "lon=-100", see details). Current region is 360.00 deg. wide by 180.00 deg. (12420.00 miles) high.

Preset Coordinates:

- Globe, USA, Alaska, Hawaii, San Francisco Bay, United Kingdom

FIGURE 3.8 Xerox PARC Map Viewer. (Courtesy of PARC, © Palo Alto Research Center Incorporated.)

to respond, and by the late 1990s, all the major GIS vendors had Web-based map viewers and servers. The creation of the Web also made it far easier for data to be taken from multiple sources and combined, a process that has become known as a "mash-up" or "mashup" (a term borrowed from the music industry). If the data was related by location, it could be displayed against a map, which is sometimes known as a "map mashup." All these applications, however, had a fairly simple structure: locate the thing of interest using either a map or addressed based search, display the results on a map, and enable simple panning and zooming, with perhaps the ability to follow a link on the map to find out more information. They did not involve any complex GIS analysis. GI had made it to the Web, but GIS had not, or at least not often.

3.5.1.4 Spatial Data Infrastructures

In many respects, this was the environment that OGC found itself in when developing its standards. The desire was to maximize the interoperability of GI, and as the Web was the natural place to exchange GI, it seemed logical that its standards should support the Web. Governments of many countries began to develop ideas around spatial data infrastructures (SDIs) in response to the development of standards and the spread of GI on the Web (even if it was fairly simple map views). An SDI is essentially a means to exchange and exploit GI using standardized methods—OGC standards being the natural choice. SDIs can be traced almost as far back as the first Web-based map applications. In 1993, the Mapping Science Committee of the U.S. National Research Council proposed a National Geospatial Data Framework.

This proposal had the full backing of President Clinton and recognized that GI needed to be part of a national infrastructure (although why just GI was singled out is unclear). All the ideas for an SDI were there, as was the political will; only the technology and the standards to make it happen were missing.

The United States was not alone, and other countries, including many in Europe, also began developing formal SDIs, as did emerging economies such as India. The European Union is currently developing INSPIRE (EU Parliament Directive 2007/2/EC), a European-wide SDI for environmental data.

SDIs are standards based and model spatial data around maps and features. Features conform to the OGC Simple Feature Model and are encoded for transport using GML. Mapping (essentially raster images) is served using WMS-compliant servers and features via WFS servers. Most SDIs also try to define the data (always government data) covered by the SDI and do so as a series of GIS style layers. Thus, most SDIs include a transport layer, a land use layer, a hydrology layer, and so on. As the standards and technology have developed, so has it become easier to achieve the aims of an SDI. However, this has also exposed other difficulties that SDIs have not fully resolved. The most significant of these have been the lack of tools to support data integration, the difficulties of addressing the semantic differences between datasets, and organizational inertia. Governments have responded to semantic differences by attempting to construct consensual standards that agree on a particular set of semantics. Such attempts tend to be long and drawn out and, as we shall see, are not in line with the principles of the Semantic Web or indeed the nature of the Web as a whole.

3.5.2   GI: A HISTORY OF THE WEB AND SPATIAL COINCIDENCE

The growth of GI within a professional context is not the only story to tell. GI, particularly in the last few years, has been significantly influenced by an amateur perspective. Two things have enabled this to happen: the establishment of Web-based mapping services that provided a resource to be used by the general population and the development of Web 2.0, where Web users became authors as well as readers.

As has been mentioned, it became easier to integrate mapping applications into Web sites, but initially this was usually hosted by an organization that placed mapping on its Web site. Web users then passively accessed the site to view the mapping and associated information. This was after all very much the way we understood the Web: The Web was a place where organizations published and the general public consumed. In the early to mid-2000s, this simple setup gave way to increased complexity as it became easier for ordinary people to publish data. The key development was the provision of tools on Web sites that allowed the user to update the site and indeed to create the content. One of the earliest examples was the comments and reviews on Amazon.com, but there are many other examples, such as wikis, where users collaboratively edit content; blogs, where individual users chronicle their own thoughts; social bookmarking sites like Delicious and Digg; or tools like Really Simple Syndication (RSS), which allows feeds from multiple sites to be collated in one place. This trend toward user-generated content became known as Web 2.0. The use of the Web 2.0 label indicated that it was felt that this advance really was

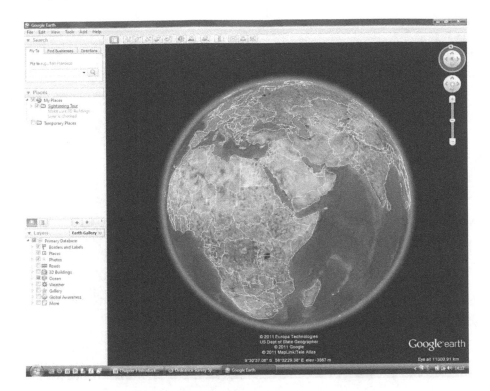

FIGURE 3.9 The iconic Google Earth.

significant, as indeed it was. The result has been to establish some incredibly influential Web-based resources such as Wikipedia, and social networking sites such as Facebook and Twitter. Other successful sites act as libraries and hosting services for community-generated material, very well-known examples being the photographic site Flickr and the video hosting site YouTube.

None of these sites is free from criticism, especially Wikipedia and Twitter, but all the sites have acquired user bases of many millions of users. So, whatever the criticisms may be, the sites are offering something that people want.

Before user-generated GI could be plotted on a Web site against map data, the technical solutions had to be a bit more sophisticated, and the value of map-based information needed to be more widely appreciated. The latter happened with the launch of resources such as Google Maps in 2005 along with Google Earth (Figure 3.9), a little later in the same year. Google Earth in particular did three things that ensured not only its success but also almost a craze for its use for a year or two following its launch. First, with its novel method of viewing, with the user zooming in from an image of Earth's globe to the area of interest, and this view being constructed from satellite and air photos, the application was not only visually appealing but also provided a more dynamic user experience. Second, it offered a very compelling user experience; third, it was free to use.

What Google Earth did, more than any other application at the time, was to make GI, and mapping in particular, something that people wanted to use, even if just to

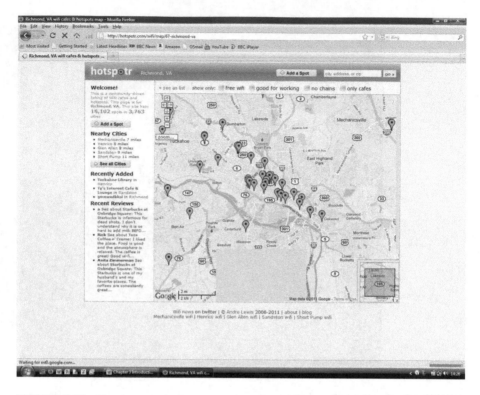

FIGURE 3.10 Hotsptr: a typical mash-up using Google Maps; this one indicates Wi-Fi hot spots (mostly) in the United States.

see what their house looked like from space: It brought mapping to the masses. For a short time at least, it was the talk of the Web. Mapping had been published by companies such as Multimap, as well as by the main search engine providers—Google, Yahoo!, and Microsoft—before the creation of Google Earth, but it was not until Google Earth made mapping cool that the general public began to become really aware of online maps. Within quite a short time, the popularity of mapping, combined with the publication of mapping resource Application Programming Interfaces (APIs) by the search engine providers, enabled users not only to view existing data but also to add their own layers of information. Mapping and GI joined Web 2.0, resulting in the creation of a large number of mashups. Here, the emphasis was often on either taking data from publically available sources or using data generated by the community and displaying it against a map backdrop: Hotspotr (Figure 3.10) is a typical example that shows the location of Wi-Fi hot spots in various cities around the world, although mostly in the United States.

Another factor that was helping to make the general public more aware of GI during the first decade of the twenty-first century was the rise of devices that included GPS. Although GPS devices had been available from at least the mid-1980s, it was not until 2000 onward that applications such as car-based satellite navigation really took off; toward the end of the decade, GPS began to appear as a feature on most smart

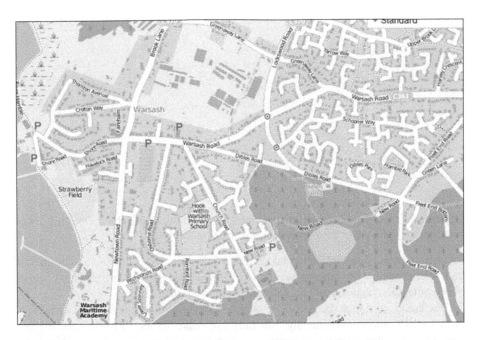

FIGURE 3.11   OpenStreetMap (2010): Warsash in Hampshire, England. (© OpenStreetMap contributors, CC BY-SA. http://www.openstreetmap.org/copyright)

phones. The last meant that location-aware applications could be developed and GI and information about a user's current location delivered to the user on the move. The result has meant that social networking applications such as Twitter can be more location aware, and users are able say where they are "tweeting" from. Users also started to geocode their content; for example, many Flickr photographs are now geocoded.

Geographic mashups exploit the availability of detailed and free mapping. But, free mapping was a problem. Where good quality and detailed mapping existed, such as in Great Britain, as provided by Ordnance Survey, it was not free; in countries such as the United States, where government mapping was free, it tended not to be as detailed or as current as in countries where there was a fee charged for map content. In the United Kingdom, a project called OpenStreetMap arose out of this frustration; the aim was to create detailed mapping by voluntary means. Formed in 2004, OpenStreetMap has since attracted tens of thousands of members not only in the United Kingdom but also worldwide. The map is created by members either submitting GPS tracks of where they have been or through digitizing Yahoo! Imagery and other free sources (Figure 3.11). OpenStreetMap was not the first community-based project to aim at capturing GI data: GeoNames, for example, is attempting to construct a world gazetteer of places, and of course many of the entries in Wikipedia (itself an inspiration for OpenStreetMap) are geographic in nature. The availability of GPS combined with smart phones has also generated new types of social applications, such as Foursquare, which is specifically tailored to providing friends with location information.

All this is not happening just within one community of people; rather, it is many different communities, each with their own aims, motivations, and preferences. But,

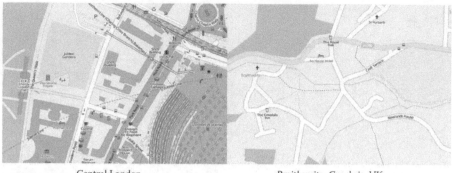

Central London   Braithwaite, Cumbria, UK

FIGURE 3.12   OpenStreetMap: London and Braithwaite in England. (© OpenStreetMap contributors, CC BY-SA. http://www.openstreetmap.org/copyright)

there is one shared expectation: that all this data is free and is available to be shared and recombined with other data. Indeed, this is the guiding principle on which Wikipedia, GeoNames, and OpenStreetMap were founded.

The creation of such datasets is not without issues. Collection can be very biased, both socially and geographically. Not only are contributors typically middle-class and male, but also strong evidence exists that the majority of work is conducted by a minority of the membership. Geograph (http://www.geograph.co.uk) is a Web site that allows contributors to send in a photograph of a landscape with the intention of providing at least one photograph for each square kilometer of the United Kingdom. Based on data for 2008 obtained from Geograph, the authors noted that at that time 1,738,728 photographs had been submitted by 9746 contributors, but that the top 20 contributors had submitted 25% of all the photos, and the top 100 had submitted 50%. Similarly, in analysis of tagging of OpenStreetMap for England, of the 3332 unique editors, 231 contributed 71% of all edits, and 39 "superusers" contributed 39% (Mooney and Corcoran, 2011). This is not atypical for crowd-sourced data. Furthermore, contributions are normally skewed heavily in favor of urban areas (although not so much for Geograph); contrast OpenStreetMap for London and that of Braithwaite (a much more rural location) in Figure 3.12.

The content can also vary significantly in detail depending on who collected it. And, there are questions over the longer-term sustainability of these projects: People to tend to like to fill blank canvasses rather than update what is already there; what happens when the next cool thing comes along? But nonetheless, we should not underestimate the achievements of these initiatives; if nothing else, they indicate the willingness of many members of the public to contribute in this way and on many different topic areas, not just geography. It also shows that there is a need for GI to make a lot of these applications work, whether GI is at the center or not.

3.5.2.1   Open Government Data

Last, turning to the government sphere, recent trends have led to the freeing of much government data, especially in the United States and United Kingdom. In the United States, where federal data is already available on request, the challenge has been to publish the data such that it is easily accessible and meaningful. The United Kingdom not only faced this issue, but given that much government data were not freely available but charged for at commercial rates, also had the additional challenge of how to make some of this data free at the point of use. This has required a change in government policy, resulting in the commercial model being retained and the government purchasing data from the appropriate government agency that is then made freely available to the end users. However, even though significant amounts of government GI are now freely available and accessible around the world, this has exposed the issue of semantics: What does the data really mean? It has also resulted in a plethora of different formats being published, making it difficult for people to process the data. So, although data is now becoming more widely and freely available, it is not always in a form that is easily digested or understood.

3.5.3 THE FORMAL AND INFORMAL TOGETHER

If we bring these two stories together, we should get a good picture of where GI finds itself at the beginning of the second decade of the twenty-first century. Crudely, we have two different communities, one professional, the other amateur. But, in reality the picture is much more complex: Both the professional and amateur worlds are not single communities, and membership is not exclusive; the GI professional by day may spend leisure time constructing mashups or updating OpenStreetMap in the evening. Perhaps it is best to first look at usage rather than communities.

What we have is at least three different uses of GI:

1. GI used within traditional GIS, back office and very specialized
2. GI used to integrate data as a mashup, or to produce integrated datasets that may have location as an aim or enabler
3. The capture and creation of GI resources such as digital mapping and gazetteers and the infrastructure to deliver them

The last use is indeed a use, although it may not appear to be so at first glance, but in most cases, other GI is used in the creation process. At this stage, it is also worth asking, Who are the users? Broadly, they fall into two groups: those that are the final end users interested in the information imparted and those who deal with the data to deliver it to the end user. Again, the situation is blurred as membership is not mutually exclusive. End users range from members of the general public wanting route directions from a satellite navigation system and those wanting to obtain information about somewhere to be visited, to professionals attempting to gauge the impact of a potential flood event or wishing to detect fraud. The intermediate users are in many senses not really users as such; they tend to be skilled in information technology (IT) (whether as professionals or amateurs) and manipulate the GI to enable the end services to be provided. But, at present these people are very influential in the direction of GI, in many cases more so than the end users. This is because they have more control over the technology. Of course, being an intermediate user does not preclude also being an end user, and there are many people who are, but for the most part their requirements and viewpoints are different. This can cause problems as it may well be that the technologists do not always understand the needs of the real end user. This is no different from many other industries (and we use industry in the widest sense), but neither does it excuse it. It is equally important to recognize these intermediate users and the value that they bring, something that applies not only to intermediate users but also to those responsible for creating the raw GI and the infrastructures.

If we revisit our three main uses, then we can also see some general patterns when we map these uses to the types of people carrying out the use cases. Those who use GIS tend to be professionals and function centric; that is, because they are interested in performing often-complex GI analysis, their interests are driven by the functionality that can be supported and the quality of the data that they are using. They mainly come from government, large industry, specialist consultancies, and software houses. This group would strongly associate itself with GI and recognize it as a distinct concept.

Those who use GI to integrate data may fall into a number of different groups and will include many from the GI community; those with an interest in non-geometry-related activities such as commercial companies that link data through address matching; and those who wish to build mashups, whether governmental, commercial, or private. These groups are quite varied in nature; some may closely associate with GIS, and others may barely recognize GI as a distinct thing.

Those creating GI and the related infrastructures tend to fall into distinct groups:

- Government agencies and usually large commercial companies interested in creating high-quality GI assets; most typically, these are national or regional government agencies collecting or creating information on mapping, environmental, transportation, and so on, along with commercial companies such as NAVTECH. In this group, there is a clear interest in quality; therefore, standards are seen as important. This group is nevertheless more quality centric than standards centric.
- Those interested in creating SDIs or similar infrastructures. This group is dominated by governments and is particularly standards centric. Here, standards are seen as the foundation of an infrastructure through the mandate of particular ways of structuring and communicating data. Standards are employed to achieve interoperability by removing diversity and standardizing both structure (syntax) and meaning (semantics). Outside this group, there are those such as Internet search companies like Google and Microsoft interested in supporting services for their users. Here, the driver is not standards but the need to attract users; hence, innovation and usability become the main drivers.
- Last, there are those, such as the authors of OpenStreetMap, interested in creating freely available GI of a quality that makes it suitable for use in mashups. The motivations for this group are many and varied and range from altruism to a general distrust of data from commercial or governmental sources. Indeed, there is frequently a tension between contributors of volunteered geographic information (VGI) and those working for government mapping agencies and other parts of government responsible for the creation of GI, with each mistrusting the motives and abilities of the other.

But, the reader should be reminded that these classifications are crude at best and are more tendencies than classifications. Also, these groupings will obscure the richness of diversity that is the true picture. Therefore, they should be treated as indicative of general trends rather than precise classifications.

This section began with the statement: "If we bring these two stories together, we should get a good picture of where GI finds itself at the beginning of the second decade of the twenty-first century." The operative word here is *should*. It is apparent that there is no one picture that properly embraces the story of GI. Rather, we have complex patterns of use and community. There is no one clear direction. Interaction between the various groups ranges enormously, as do the usages. But, perhaps it is this very complexity and incoherence that are the common factors. All these groups ultimately share the same problem: Using GI is often hard. "Hard" limits what can be achieved by creating demands on time, money, and required knowledge. It is hard because GI does tend to be more complex than a lot of other data, especially that which relates directly to geometry or requires spatial indexing. It is hard because it exists in so many forms, and despite its existence, only a small proportion of GI fits any standards. That so much has been achieved is to the credit of those involved with GI. At least those who deal with GIS, and who probably most strongly recognize GI as a distinct thing, are aware of these issues as they have to deal with them on a day-to-day basis, as are those working to develop standards and infrastructure to help resolve these issues. But, the largest group, those who use GI, often unconsciously, to join data or produce mashups are the ones who struggle the most and are most constrained. For these people, the struggle is with the syntactic and semantic diversity. Indeed, most mashups in truth are little more than "pins on maps," limited to display simple locations with some attribute data as content. Anything more would have required too many data translations and probably too many semantic errors, partly due to imperfectly described data.

There are also social issues:

- The mutual distrust that often exists between professional authors of GI and those involved in the creation of VGI.
- The insularity of the GI community—that body of mostly professionals that explicitly recognizes GI as a distinct type of data. Such insularity is not atypical of particularly specialist professional groups, but perhaps it is because geography offers so much potential for integrating other data it is more of an issue for GI. The more people who see GI as special (and maybe therefore not for them), the more difficult it will be to use GI as a means to link across communities and hence data (because it is within communities that different data resides).

3.6 SUMMARY

If the reader finished Chapter 2 with a largely coherent picture of Linked Data and the Semantic Web, the same probably cannot be said of GI at the end of Chapter 3. GI is marked by diversity in types of data and communities of users and creators. There are those who have a strong sense of GI as something different, and many who would not really recognize that they were dealing with GI.

From a traditional GIS perspective, geometry is the main driver, and the world is divided into different layers of information, each layer representing a particular type. Standards are also geometry centric and support the concept not only of layers but also of features: OO representations of real-world things where information about the same object is collected together. Other views also exist; these are focused less on geometry and coordinates and more on relationships, whether these are topological, mereological, or lexical. Standards for representing GI have been developed, but these are largely restricted to use by a subset of the GI community, are little known outside this community, and may even serve to help enforce its isolation. GML is an example of this: used by governments, mandated by policy, but not widely known or applied by those who use GI in its widest sense.

If we were to simplify things and if we just separated the uses of GI into two groups, however imperfect this may be, then GI can be broadly split into

1. GI for analysis. Here, the emphasis is to generate new information by performing special spatial operations between different datasets.
2. GI for integration. Here, the purpose is to combine different data, either as an end in itself or as a prerequisite to performing spatial analysis that typically involves multiple data sources.

Our focus in this book is to bring together Linked Data and GI. It is therefore natural that we concentrate on the latter use of GI for integration, as Linked Data has much to offer in this area, and it addresses the widest audience. In the next chapter, we start to examine at a high level how the ideas behind the Semantic Web and Linked Data require us to think differently about GI (and indeed about information and data generally). This next chapter is thus extremely important as through the introduction of this different way of thinking about data we provide a foundation for the rest of the book. Following from Chapter 4, the book introduces the more technical details behind Linked Data and Semantic Web ontologies and shows how these can be applied to GI.

NOTES

1. Mereology describes whole–part relationships.
2. *Real-world objects* is a GI term used to refer not only to physical geographic features such as buildings, roads, woods, and rivers but also less-tangible things such as an administrative area such as a state or county or health district and even something that may only exist for specific times in the same place, such as a Monday farmers' market.
3. While it is true that European directions more frequently include landmarks for reference than, say, a set of directions given by an American, it is still notable that they are wholly absent from the machine-generated directions.
4. In OGC terms, a feature is a digital representation of a real-world object, reflecting a single perspective of that object and containing some explicit locational information.
5. Strictly, it could be argued that KML also originated with the GI community. KML was created by Keyhole Incorporated to enable geospatial visualization but was picked up by many casual Web developers needing to represent mapping in their applications. This success led Keyhole to be absorbed by Google in 2004.

4 Geographic Information in an Open World

4.1 INTRODUCTION

The previous chapters introduced the ideas behind the Semantic Web and Linked Data. We also introduced Geographic Information (GI), although it paints a less-coherent picture, reflecting less consistency and greater diversity in this area. This chapter represents an initial meeting of the two—an introduction—and it can be considered an anchor for the rest of the book. It sets out the general technical principles required to represent GI on the Semantic Web, but before addressing this area, we cover two other matters. We describe the approach that must be taken, contrasted against the prevalent approach to mainstream GI today; and identify where the Semantic Web can be used to best effect. There are significant, though sometimes subtle, differences between the traditional GI handling of data and the strategies of the Semantic Web, so it is important to clearly state how we must tackle GI on the Semantic Web. The need to identify where the Semantic Web can be used to best effect reflects the nature of all technologies and methodologies: They are not panaceas, they not only have strengths but also weaknesses, and it is as important to recognize if a technology is inappropriate as it is to know when it can be used to advantage.

If the benefits of incorporating GI into the Semantic Web are to be fully realized, then it is not sufficient just to reimplement what we already have available in traditional systems. Instead, we need to stand back and look at how we should address the issue afresh. It is from this perspective that we will begin, by comparing the underlying principles that are currently associated with both the Semantic Web and GI. Only then do we move onto the technical aspects, laying down the foundations on which the rest of the book is built. From the technical perspective, we also take a route, reflected in the structure of this book, that mimics the way that we believe the Semantic Web will grow, with an initial emphasis on the publication of data as Linked Data before rich description is added through the development and publication of ontologies that semantically describe the data.

4.2 PRINCIPLES

4.2.1 SEMANTIC WEB

The Semantic Web is founded on what is known as the open world assumption. The open world assumption has both a strict formal definition and a looser projection of this definition from the abstract world of theory to the real world of the Web (if the Web can ever be described as real). Formally, the open world assumption is that the truth of a statement is independent on whether it is known to be true or false by any single person. That one person does not know a fact to be true does not mean that it is false. This may seem to be intuitively obvious, but it is not the supposition made by most databases. Here, the closed world assumption applies: It is presumed that a statement is false unless it is known to be true in the database. This strategy is valid in that most databases are built to be a closed solution: They assume their universe comprises only the things of interest to the database; therefore, if a fact is not stated in the database, it is not true, or at least not true within the scope (narrow worldview) of the database. The scope of the database therefore defines the limits of knowledge. This is a useful assumption to make for most databases as it means that data can be checked for compliance to the data model. But, it is not valid for the Web as the scope of the Web is without limits. The Web is also decentralized and has no global data model; it grows organically, and therefore it is not possible *a priori* to know how new information will fit, and it is certainly inevitable for there to be conflict between the data on the Web. Things could not be more different from the orderly, defined, and controlled world of a database.

On the Web, the open world assumption holds sway. This is an extremely important observation. It means that for the Semantic Web there can be no single worldview, no God's eye view, no one model that fits all. It also means that *facts* may conflict with each other, and different people may resolve these conflicts differently depending on their own worldviews.

The degree of agreement on the Semantic Web is at the lowest level possible, with the standard languages (Resource Description Framework [RDF] and Web Ontology Language [OWL]) used to encode knowledge. Beyond this, and within the knowledge encoded in OWL, the open world assumption applies. To think sensibly about the Semantic Web, and indeed the Web as a whole, one has to assume that one's own knowledge is incomplete, and that publishing on the Web contributes to a greater pool of knowledge.

The Semantic Web also provides an explicit and formal description of the data through the use of ontologies. An important aspect of this is that the ontology is independent of both the data and any application code. This makes the ontology very visible and provides greater flexibility because ontologies are easy to share and reuse or adapt for specific situations.

4.2.2 Geographic Information

With a field as diverse as GI, there are no comparable founding principles as there are with the Semantic Web. Whereas the Semantic Web is built on the open world assumption, there is no such positive assertion for GI. However, and accepting the caveat that because of this diversity there can be no universal statements, it is fair to say that for the most part GI is founded on the closed world assumption. This is a reasonable statement given that the technologies most commonly used to implement GI today, Geographic Information Systems (GIS) and relational databases, are implemented based on the closed world assumption. Furthermore, this assumption has also been carried forward into Open Geospatial Consortium (OGC) standards. Consider the following short paragraph describing an OGC feature: "Any feature may also have a number of **attributes**: Spatial, Temporal, Quality, Location, Metadata, Thematic. A feature is not defined in terms of a single geometry, but rather as a conceptually meaningful object within a particular information or application community, one or more of the feature's properties may be geometric" (OGC, 2011).

The OGC feature is a basic component of OGC standards and is an abstract representation of some *aspects of* a geographic feature in the real world. The quotation contains a number of key aspects about a feature. First, it indicates that a feature is a data structure that is restricted to a single information or application community—a closed world with a strict and limited scope. Second, the feature wraps a number of attributes as a single tightly bound package. It defines *all* the attributes that the feature may have; some can be optional, but it cannot have additional attributes that are not described by the feature's definition. Again, this is closed world. An OGC feature represents a data abstraction of a real-world object rather than the object itself. Each OGC feature has a unique identifier, the Feature Identifier or FID, and this identifier is associated with the OGC feature, not the real-world object it describes. This is in stark contrast to a Semantic Web identifier (Uniform Resource Identifier, URI) that explicitly represents the real-world object, is not bound to a particular community, and identifies an object that may be incomplete in data terms. The difference in representation between abstract and real features is important and often causes much confusion, as it can be quite a subtle difference. Thinking about something as an abstract representation is good when there has to be agreement between parties about what is being referred and works well with internalized models; identity is applied to this abstract data. One problem with this view is that ultimately the abstract has to be grounded in the real, so in certain circumstances it can be seen as an unnecessary complication. A viewpoint of the item of interest as the real-world object is better suited to information exchange and to an open Web. Here, identity applies to the real thing.[1] It is not that the Semantic Web does not have abstract views; it does, and these are represented in ontologies. The difference is more about how they are presented. In the traditional database or object-oriented (OO) view of the world, the abstract is bound exactly to the data. The abstract model represents the information that a particular application requires, and this dictates the data structure that is needed to hold that information. In the Semantic Web, the abstract model represents a minimum classification for something and is held in an ontology, separate from the data. Therefore, it is possible to represent more about a real-world object than may be contained in any one abstract model. For example, an ontology may state that the minimum requirement for a school is that it has a building with the purpose of providing education. In the Semantic Web, this does not prevent us from holding additional data about any specific school, perhaps saying it also has a car park and sports fields. The abstract model defines minimum membership for the class "School"; the real world is more complex, as is the Semantic Web data. In contrast, the traditional abstract model provides a complete application definition, so if the application needs to know that schools can have car parks, it had better say so.

In a pre-Web world or when an organization is interested only in an internal solution (or one that is exposed to parties that all share the same model), the closed world assumption is perfectly valid. It is also efficient because it is able to enforce strict compliance; data that does not conform to the model can be rejected and the quality and integrity of the database ensured. The system also does not have to worry about anything declared as out of scope. But, the closed world assumption does not work well in eclectic environments simply because different models coexist; therefore, assumptions that may be valid for one model may be invalid or irrelevant for another. The use of GI itself is becoming more eclectic as its uses broaden, particularly through the informal application of GI that marks much of the growth of Web-based services.

Closed applications require standards that enforce a single model in place of the many that may have previously existed to make it easier for organizations to share data. Spatial data infrastructures (SDIs) can be seen as an embodiment of the International Organization for Standardization (ISO) and OGC standards that exist for the GI community; they provide a physical framework for the exchange of GI through shared and standardized models and transport mechanisms. But, such solutions are not scalable to the Web as a whole, or rather to the Web *community* as a whole. It is simply impossible to enforce a single world model. The problem is of course that the GI community's view is not the only view, and even within the GI community there are many different views. Other communities have their own standards and frameworks to exchange data. And, the issue is further complicated by the fact that an organization is rarely a member of a single community. As a result, there are plenty of occasions when an organization has to take data structured according to one set of standards and translate it (usually in conjunction with other data) to conform to another set of standards. An insurance company concerned with exchanging property-based data may have to transport data in both GI and insurance industry standards and will therefore require internal mechanisms to translate between these standards. The translation process itself is normally done within the company, and the translated data is unavailable to any other organization facing the same problem. While this hiding of data may protect a commercial interest, it is inefficient and is likely to be less attractive in noncompetitive situations such as government or other data in the public sphere.

The need to enforce standards to efficiently exchange information within a community has had the unfortunate side effect of reinforcing the insularity of that community. Another unintended victim of this approach is the end user; the compromises made to develop the necessary standards are often made at the expense of usability of the data for the end user. Computational and databasing efficiencies, which are often important drivers of standards, do not necessarily provide the best end-user solutions in terms of data content or comprehension. The availability of data that without standards might not have been otherwise accessible does mitigate inconveniences in usability. Nonetheless, standards do not always provide end users with an optimal solution.

Like many other communities, the GI community has a tendency to think of itself as the center of the universe, however unconsciously this happens. From this perspective, GI is king; everything else is secondary. It is undeniable that in many, many instances geography is an important element, but it is not always so, and an over-inflation of this importance can be distorting. Consider the following: An oft-stated *fact* within the GI community is that 80% of all organizational data is geographic. This statement frequently occurs in publications and talks given by those in the

GI community and has been so for many years. We have seen in the previous chapter that geography does indeed occur in many areas, and it is this frequency of occurrence that makes GI a useful vector for data integration. But 80%? Although the statement is often made, it is never accompanied by any reference to substantiate it; it is almost as if because it has been around for so long it is unchallengeable, and in any case logically it must be right. Well, maybe. In fact, some have questioned this and investigated its origins. The statement can be traced to the following: "A 1986 brochure published by the Municipality of Burnaby, British Columbia, reported the results of a needs analysis for an urban Geographic Information System (GIS) in that municipality: eighty to ninety percent of all the information collected and used was related to geography" (Huxhold, 1991).[2] So, it appears that this statement only refers to data found within a local authority where one would expect there to be a significant geographic interest. It cannot be extrapolated to cover all organizations. Further, the original source was a brochure—perhaps not the most reliable source of information and almost certainly not subject to peer review. So, how factual is this "fact" after all? The best that can be said is that it is unproven, unlikely to be correct, and most likely to be an overestimate. Now, the point is not that GI is unimportant, but that this is indicative of the GI community overestimating its importance, and that this reinforces a GI-centric view of the world. It is perhaps also enlightening that a community could be so confident in its own self-importance that one of the most frequently repeated statements was not questioned for almost 20 years. Lest we are too hard on the GI community, it is fair to say that this introverted nature is true of most professional communities to a greater or lesser extent. In fact, for these communities *working within themselves*, it is also a perfectly reasonable viewpoint. If you are only concerned with exchanging information with like-minded people, then the specialism that binds you together is the center of your closed world and enables local standards to be agreed through a shared understanding.

This has resulted in the emergence of many different communities that are learning to communicate better within themselves (through local standards) but struggle when exchanging data with others. If only other communities would realize that they should talk to us using our standards. It is oft said that the problem with standards is that there are so many to choose from, but we would argue that the fundamental problem is that most standards only work well within a particular community. Certain standards can have a truly global reach: The metric measurement system is one such obvious example, although even here there is still some resistance in certain Anglo-Saxon societies. But, more often than not standards are specialized to a single community of users.

Last, it is worth remembering that while a formal GI community exists and operates with agreed standards (especially within government), increasingly GI is being used in a more informal sense by those who would not consider themselves members of the GI community and are probably unaware of its existence or standards. These people will use GI as a means to an end and are more and more doing so in the form of Web applications and services. For these people as much as those using more formal methods, the Semantic Web and Linked Data offer the potential to better exploit their information resources.

The Semantic Web will not completely solve all problems related to data exchange and recombination, but it does offer ways to reduce and better manage them. The Semantic Web also cannot help in all circumstances; it is better suited to certain types of GI-related problems than others. Equally, the specialist knowledge developed by the individual communities must not be lost but preserved in the context of the Semantic Web. The next section identifies where the Semantic Web can help and where it cannot.

4.3   APPLYING THE SEMANTIC WEB TO GI

In general terms, GI has two major uses. GI is an essential element in data analyses for which location and space are important elements (spatial analysis), and it is also used to combine data through location. In almost every case, geographic data analysis cannot be performed without the ability to combine data through location, and elements of the former may also be required to assert that two entities are collocated. The situation therefore appears to be circular, but can be resolved.

The need to perform spatial analysis was central to the emergence of a GI community and the creation of GIS. As has already been discussed, GIS mostly operates through arithmetic calculation performed on explicit geometry, not an area where the Semantic Web is strong. The emergence of new standards such as GeoSPARQL that provide some simple geometrical operations for data held within the Semantic Web do enable simple spatial analysis to be performed based on geometry, but this is still a long way from the sophistication provided by a GIS. Therefore, the Semantic Web is limited to simple spatial analysis plus the ability to perform some topological analysis. On the face of it, this is pretty damning regarding the usefulness of the Semantic Web to the GI Community or anyone wishing to perform some spatial analysis.

4.3.1   EXAMPLE

Let us suppose the following fictitious situation: Along the banks of a river, it is noticed that an invading plant occurs unevenly, and the spread is also uneven. It is postulated that the plant thrives where the surrounding countryside has a cover of unimproved grass and where there have been sightings of a now-rare spider. Away from the riverbanks, there appears to be no correlation between spider and plant. It is decided to perform a spatial analysis to confirm or dismiss this theory.

The actions that are necessary to complete this task are to obtain the appropriate datasets, load these into GIS, perform the necessary calculations, and present the results. Expanding each of these tasks, we can gain a better understanding of the nature of individual processes.

4.3.1.1   Obtain Appropriate Datasets

If we assume that the analyst does not have immediate access to the data, then a search for the relevant datasets is first required. At present, although a few metadata sites exist, the majority of data is not discoverable through directories, and the process will require a degree of investigation. Having found a prospective dataset through cunning sleuth work, our analyst must confirm that the dataset does indeed include suitable data. This process will examine a number of different aspects of the data and will include understanding the content and the level of detail (scale) that the data represents: If the spider-sighting dataset only contains sightings to the nearest 1 km, is this sufficiently precise? Understanding the nature of the content is a semantic exercise. For example, if the data is of spider sightings, then is the species of interest included, and can it be discriminated from the other spider species? Similarly, if the data describes land cover, then is the right land cover included, and if it is, is the classification system that is used understood, that is, is unimproved grass as represented in the data the same as the analyst's expectation and requirement? This last point can be quite subtle. What it is really asking is whether there is (semantic) agreement between what the data collector and the analyst understand to be unimproved grass. Often, such terms are used in similar but not identical ways, and if the two definitions are not identical, the analyst then has to ask whether this will have a significant impact on the analysis. To answer these questions, the analyst will be reliant on the quality of the documentation and may even have to directly contact those responsible for the production of the data. There are of course other considerations when obtaining the data, including such things as permissions and whether the data is freely available or commercial in nature. But, for our discussion the important aspects are the locational and semantic appropriateness of the data.

4.3.1.2   Load the Data

Having obtained the data, the next step is to load the data into the GIS for analysis. If the analyst is lucky, the data will already be in a format that can be loaded directly into the GIS. When dealing with inherently spatial datasets such as the ones that are likely to be required for this exercise, there is a reasonable likelihood that this will be the case, or the data will be in a format that the GIS is able to accept through a supported translation process. But, it is not always so. Let us suppose for now that although most of the data can be loaded in a straightforward manner, the invasive species data has been created using an unfamiliar data format. Now, it is necessary for analysts to understand this format and to convert the data to a form that is acceptable to their GIS. Depending on how well the data is documented, this can be a fairly challenging, though tedious, task.

4.3.1.3   Conduct the Spatial Analysis

The final stage requires the analyst to process the data to test the hypothesis. This can be achieved by executing the following spatial query: "Locate all invasive plant sites that are on riverbanks adjacent to unimproved grass and near sightings of the spider."

This, of course, requires the query to be a little more specific: *Adjacent* may be interpreted as "within 20 m" to allow for areas where the unimproved grassland may be close to but not actually physically next to the riverbank. And, *near* may be interpreted as within 50 m to allow for the likely travel distance of the spider. The query is then resolved by the GIS, creating buffers of 20 and 50 m around the river and the spider-sighting locations and then selecting those invasive species locations that are contained within overlaps between the two types of buffered areas. Similar queries can then be executed to find plant locations that do not fit the original criteria, and by comparing the different sets of results, a conclusion can be reached that either supports or disproves the hypothesis.

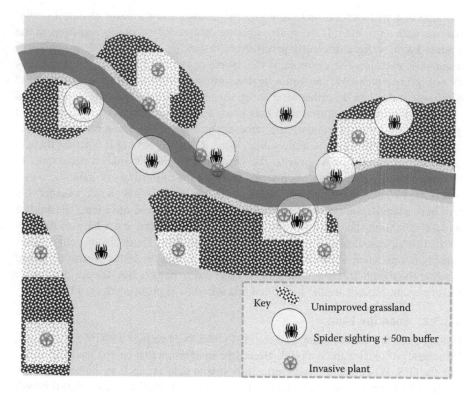

FIGURE 4.1 Distribution of Spiders and Plant in unimproved grassland within 20 m of the river.

4.3.1.4 Observations and Discussion

In this example, the integration between the datasets is loose and relies entirely on the spatial calculations made to determine the relationship between the plant locations and those of unimproved grass and the rare spider. These relationships are therefore not explicit and calculated arithmetically using the underlying geometry associated with the various locations (riverbank, unimproved land, plant locations, and spider sightings). When selecting the datasets, there was no standard way to establish whether a dataset actually contains the right data. This is a semantic issue: Was the understanding of unimproved grass the same for both producer and analyst? And, it was also necessary to understand the detailed structure of datasets and restructure them to a suitable format to be loaded into a GIS. The results of the GIS analysis are shown in Figure 4.1.

How much easier might things be if the problem was approached in an environment where the Semantic Web was well established, with many datasets available as Linked Data and semantically described through ontologies? The answer is that some stages will be easier and some not at all or only a little. Acquiring data will become easier. Ideally, it would be possible to query the Semantic Web to look for appropriate datasets and then use the detail of the ontological description to determine the semantic agreement between the user's requirements and those of the data. It would also not be necessary to convert any data, as in this idealized world all data would be held as Linked Data, and a GIS would be capable of importing this directly. If we assume the analysis itself is performed by GeoSPARQL, then it could be done in this simple case but almost certainly less efficiently than the GIS, and even this relatively simple case is more or less at the limit of GeoSPARQL's capabilities. For tasks any more complex, it is unlikely that GeoSPARQL would suffice. And, this example is not only a fairly simple example of spatial analysis, but also in many respects a simplification of what would be required to perform a more robust analysis. In reality, it is more likely that the vegetation would graduate from one area to the next, and this would require spatial modeling and analysis techniques beyond the capabilities of GeoSPARQL. Put simply, the Semantic Web is not terribly good at arithmetical analysis, and without GeoSPARQL, the Semantic Web is even more limited in this respect. The Semantic Web is much more able to process explicit relationships than perform arithmetic operations, so if spatial relationships in the data were presented as explicit associations, then it would be possible to obtain the correct results. This still means that all the relationships need to be precomputed and stored, quite unrealistic given the sheer number of possible spatial relationships that exist between objects; we would in effect require anything that can be located on Earth to be related to all others. Even when considering the initial task of dataset discovery, it is unlikely that this process would be completely automatic. An ontology cannot by its very nature be a complete description of the data or what the data represents. Therefore, it is likely that the analyst would still want to intervene to confirm the semantics. Despite these limitations, the Semantic Web will nonetheless still make this process significantly easier and less prone to error as the data descriptions would all be explicitly defined in a standard machine-readable language. The Semantic Web is also good at describing the relationships that data has to other data, at representing the data in a universally uniform manner, and at enabling inference to be made about the data. It is not a system designed to perform specialist arithmetical or statistical computation. Thus, although extensions to the Semantic Web such as GeoSPARQL enable a limited amount of spatial analysis, in many cases the use of a GIS as an analysis engine will still be required.

The Semantic Web will not replace GIS; each must be used in the most appropriate manner. GIS is there to perform specialist analysis; the Semantic Web is more about organizing data for maximum reuse.

4.3.2 Topological Relationships

Spatial relationships do not solely exist as implicit geometric associations, and there are circumstances when the Semantic Web can indeed be used more effectively to perform analysis. Consider a government that releases a number of different datasets, one containing educational achievements for each school, another showing areas of social deprivation, and another showing health issues by area. The school information identifies each school by name and address; the areas of social

TABLE 4.1

School Performance Assessment

School	Road	Area	City	Postcode	Assessment
Bay	Elm Lane	Bove	Eagleton	EA1 7GG	5
Court Place	Brambles Road	Bove	Eagleton	EA1 8QT	3
Beech House	West Street	Central	Eagleton	EA2 9WW	8
St John's	Derby Road	Hasly	Eagleton	EA2 1AC	6

Deprivation Areas		Health Index	
Area	Deprivation Score	Postcode	Index
Bove	3.9	EA1 7GG	97.2
Hasly	3	..	
Central	1.2	EA1 8QT	88
Merriby	2.3	..	
		EA2 WW	93.4
		..	
		EA2 1AC	93.7

deprivation are identified by the name of the area and the health issues by postcode. We are able to relate this data using conventional means such as relational databases as there are common factors that can draw the data together. The use of GIS in this case is not necessary.

Table 4.1 shows the relational database tables for schools, health, and deprivation data. The address element of the school will enable us to associate the school both to the areas of social deprivation via the area element of the address and to health issues data through the postcode element. It is now possible to look at the correlation (without implying causation) that social deprivation and local health issues have with the school performance results; however, this can only be achieved by stating the relationships using an SQL Query; the associations that exist between the data are not explicit.

However, if the data were published as Linked Data, then an initial step in understanding these data structures and converting them to a common model would be significantly reduced, if not removed altogether, and the relationships between the various elements represented as explicit links rather than revealed indirectly through an SQL Query. By publishing the links on the Web, there is a third benefit as now these links are available to others. Table 4.2 and Figure 4.2 show how the data is represented.

In conclusion, the Semantic Web can improve the processes of locating and presenting data to a GIS for analysis, but the GIS is much better at performing arithmetic analysis. The Semantic Web is better suited to cases where the task can be accomplished through inference or graph analysis. In both examples, the representation of different datasets in the common data model of Linked Data reduces effort required to massage data so that it is suitable for analysis, and the links themselves can be made available to others, providing onward benefits.

TABLE 4.2

Individual Statements (Known as Triples) for Bay School

Bove is within Eagleton.
Bay School is within Bove.
Bay School has Postcode EA1 7GG.
Bove has Deprivation score 3.9.
EA1 7GG has Health Index 97.2.

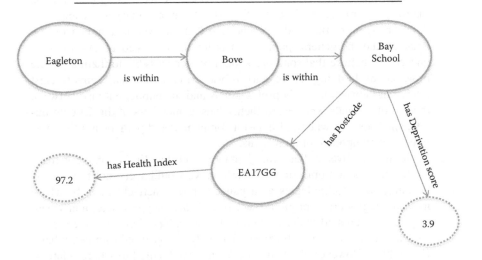

FIGURE 4.2 Linked Data graph for Bay School.

4.4 IMPORTANT OBSERVATIONS

What follows are our personal observations on designing solutions for the Semantic Web and ones that underpin the rest of this book:

1. No person has a God's eye view; this is a privilege reserved for God. The Web is a reflection of the world and a world that is heterogeneous and eclectic with diverse and often conflicting models used to describe different aspects of it. There is no one complete and encompassing model. This has a number of implications. First and foremost, the world and the Web are not geocentric—we have to adopt a view that geography is but one component of a much richer landscape. This also means that we cannot look to standards to provide global solutions; we also should not look to standards to try to engineer out *necessary* diversity.[3] We must instead look to find efficient ways to work with such diversity. This does not mean that we cannot *choose* to take a geocentric view when it is appropriate. But, when we do so we need to remember that it is not the only view, not necessarily always the most important view, and that we need to model the world in a way that enables us to adopt different views when it is sensible to do so.

2. It is important to take a user-centric approach and design for as wide an audience as possible. Here, the emphasis should be on the end users: those people who have to make decisions based on the GI they are presented. This is to differentiate between them and the intermediate users who may deal with the GI to prepare it for use, whether this is in a professional context, perhaps a GIS expert loading data into a GIS, or an amateur building a mashup. This is not to say these groups are not relevant, but that it is more important to preserve the worldview of the end user, something frequently distorted for the convenience of the intermediate user. Designing for the intermediate user is also important but should be secondary to the end user. Another aspect of this is that we believe that the Web is as much for the amateur and the general public as it is for the professional. As a consequence, we believe that solutions should be as accessible to them as to the professional. In doing so, we start to bridge the divide that seems to exist at present between not only professionals and the public, but also between those professionals who regard themselves as members of the GI community, and those for whom GI is just a component and perhaps not the most important component of their focus.

3. There are no silver bullets; the Semantic Web has limitations, but these are limitations, not objections. Therefore, we would argue not to treat the Semantic Web technologies as a panacea but to actively identify where its technologies are appropriate, where they are inappropriate, and where they may be applied under certain restrictions. The Semantic Web is not an appropriate tool for complex spatial analysis based on arithmetic; here, GIS is king. However, the Semantic Web is well suited to where relationships can be made explicit; often, it will be a more appropriate solution than a relational database where much of the semantics can only be revealed through SQL queries.

4. Scope and purpose: The problem with the Web is that it is boundless and constantly growing. This open world takes us away from the closed and small universe of the traditional database, which is nicely scoped and bounded. This can present difficulties for those first attempting to build Linked Data, and particularly ontologies, for the first time. Where does one stop? Scope creep can occur both in breadth (by adding additional concepts outside the main focus area) and depth (by introducing concepts that add unnecessary additional detail to the ontology that will never be used). Here, an important aspect is to define the scope and purpose of what you are trying to achieve; this is not too dissimilar to what one would expect to do for a conventional database solution, but there is a difference. The scope and purpose define *your* limits, not the limits of the system in its totality—the Web. This might at first sound contradictory. If you design with a particular scope and purpose in mind, how is this really different from designing a conventional database? However, given a little thought, the difference becomes clear. When you design for a database, you have complete control of everything; when you design for the Web, you only control your element: You need to think about and design for the way your element will be seen

by others and how they might wish to interact with it. Therefore, clever solutions for a database that you can hide from a user may be confusing if repeated on the Web. We expand on this in further chapters.

5. Separation between data and description: The main benefit of this is that the description is now far more visible within an ontology; it is not locked in the structure of the database, the query, or the application code. It also has the consequence that others may choose to interpret your data differently from the way you do, and likewise you may choose to interpret others' data differently. Therefore, by applying a different ontology, the data can be interpreted differently. This may seem just wrong—surely you should always use the data in the way the publisher intended? The first thing to realize is that we often already reuse data in ways not intended by the publisher. The Semantic Web just enables this use and the consequences of such use, to be more explicit. Second, you may want to use the data with the same meaning as the publisher but just need to express it differently to fit with your wider view of your world; this could be as simple as changing the vocabulary that is used or be more complicated perhaps: changing the relationship that one thing has with another. For example, a cadastral dataset might apply an address to a property as a whole, its buildings and grounds, whereas someone interested in historic buildings may wish to take that address and apply it only to the historic building within that property. The relationships are clearly different, but each is correct in the appropriate worldviews. And last, as a publisher, you cannot ever plan in advance for all the different ways in which your data might be reused.

4.5   SUMMARY

The adoption of Linked Data, and more broadly the Semantic Web, will enable the GI community to exploit the value of GI more widely and in doing so will both expand its use and enable it to better fit within the Web as a whole. However, to achieve this, the GI community needs to recognize that GI is not so much a hub as glue. On the Web, with its open world assumption, there is little place for a GI-centric view; in a minority of circumstances, this may be necessary and appropriate, but in most circumstances, it will present a barrier to the use of the data. If we view GI as glue, then we begin to understand the role it more often takes in integrating other data. Glue is rarely the focus of attention but is a component essential for binding together so many different things. And, perhaps this is how we should view GI: as an information glue to integrate different data. This view fits more naturally within the Semantic Web, where the importance of explicit links between data is so highly valued.

The Semantic Web is not the right vehicle for performing complex arithmetic calculations and will no more replace GIS than spatial databases have. From understanding this limitation comes an appreciation of what can be achieved through the use of topology and other relationships.

This chapter completes our introduction to the Semantic Web and Linked Data. The rest of the book now describes the technologies and their application to GI in more detail, starting with Linked Data in the next chapter. This chapter describes the RDF, the basic language for expressing Linked Data, and then the following chapter applies RDF to GI.

NOTES

1. There is often an assumption that assignment of identity can only be made by the owner of the real-world thing; therefore, some people have argued that it is a limiting factor of the Semantic Web. However, we can all assign our own notion of identity to something without implying ownership, so this assumption is clearly unfounded.

2. We are grateful to Dr. Muki Hakley of UCL for tracking down the origins of the statement (http://povesham.wordpress.com/2010/02/22/the-source-of-the-assertion-that-80-of-all-organisational-information-is-geographic/).

3. Diversity can be a direct reflection of the complexity of the world; this is what we would term *necessary diversity*. It can also be a consequence of different people describing the same underlying complexity in different ways; this is *unnecessary complexity*, and here standards can help to reduce the diversity. However, it is not always easy to distinguish between the two cases. If standards are necessary, they can also be represented in terms of ontology patterns, providing a new way to express old ideas.

5 The Resource Description Framework

5.1 INTRODUCTION

This chapter looks at the Resource Description Framework (RDF)–the standardized way of encoding data for the Semantic Web. It has been said that, "RDF is for the Semantic Web what HTML has been for the Web" (Yu, 2011, p. 24). Since RDF is the primary building block of Linked Data and the Semantic Web, it is important to understand how it works in terms of both its data model and its language and how to interpret the meaning of its constructs, its semantics. In addition, in this chapter we introduce some basic RDFS[1] (RDF Schema) ontologies that you are likely to come across and may find useful to reuse when publishing your own data. Chapters 6 and 7 go into more detail about how to encode GI as Linked Data and how to publish, query, and link it on the Linked Data Web, but for now we concentrate on preparing the groundwork of your RDF understanding.

5.2 RDF: THE PURPOSE

So, what exactly is RDF trying to achieve? Its aim is to be both a "conceptual framework" (a way of looking at things) and an actual language for describing Web resources independently of any particular domain or subject area. The purpose of this is to allow applications that exchange machine-understandable information on the Web to **interoperate** more easily with each other. Since we are dealing with the Web, this framework needs to be **scalable**, so that it still works with any amount of data; **flexible**, expressive enough to encode any sort of information we might need to talk about; and yet **simple**, so that it is easy for anyone to read, write, or query it.

Let us illustrate all of this with an example. The fictional Web mapping company Merea Maps offers data about Topographic Objects via a Web service API (Application Programming Interface). In the traditional Web scenario, a mobile user wants to find out which pubs are nearby. Through an application on the user's phone, an HTTP (Hypertext Transfer Protocol) request is sent to the Web service to retrieve a list of Topographic Objects, categorized as "Pub," within a certain radius of the user's current location, specified as latitude and longitude. The list of pubs returned might be something like that seen in Table 5.1.

Although the ID number may not be presented back to the user, as it is meaningless, the Web service will almost certainly internally store an identifier for each Topographic Object. Our user decides on the Isis Tavern, visits it, and thinks it is great. Now, if the user wants to publish a review of the pub, giving it five stars and letting people know about its food and beer garden, the user will encounter some

TABLE 5.1

List of Topographic Objects Returned from a Query to a Web Service

ID Number	Name	Category	Latitude	Longitude
0001	The Moon over Water	Pub	51.647396	−1.229967
0002	The Nags Head	Pub	51.596874	−1.172244
0007	The Kings Arms	Pub	51.733995	−1.231246
0012	The Isis Tavern	Pub	51.730031	−1.241712

problems. The user needs a way of putting this review on the Web, structuring the information in a way that other people and programs can easily understand, and linking it back to the original Web service's data, which included useful information like the latitude and longitude. Merea Maps cannot know in advance what extra information any Web user might like to combine with the user's data, so it cannot provide an appropriate structure to hold the data. Their data structure just is not flexible or scalable enough to add these new types of data. Also, there needs to be a clear way of expressing identity—allowing our user to say that the review is talking about the same Isis Tavern as in Merea Maps' data—so that the two sets of data are interoperable. RDF offers a way of achieving all this.

5.3   A WORD ABOUT IDENTITY

First, RDF tackles identity in true Web fashion by assigning a unique identifier to everything. To do this, it uses the Web standard Uniform Resource Identifier or URI. As we mentioned in Chapter 2, a *resource* is simply a thing of interest to us, which could be a Web page, a piece of data, or a link between pages or data items. However, in a more general sense, it can also refer to a physical object in the real world or an abstract idea or relationship. The well-known URL (Uniform Resource Locator) or "Web address" is a kind of URI, for example, http://www.example.com/myreviews.html is a URL for the "My Reviews" Web page. However, a URI is a more general term that encompasses any item on the Web or beyond it. The other sort of URI is a URN, or Uniform Resource Name, which gives the name of the resource without explaining how to locate or access it. For example, mydomain:BuildingsAndPlaces/Pub is a URN naming the category of Pubs. As you can imagine, that is not much use for us in the Web world; we need to be able to find and use our Web resources. URIs that include an explanation of how to locate the resource that they are identifying are known as "dereferenceable" URIs. For example, http://linkedgeodata.org/triplify/node175683857 is a URI for the data resource The Lamb and Flag pub in LinkedGeoData's RDF data, and you can look it up at this location. We discuss how to make your URIs dereferenceable when we talk about publishing Linked Data in Chapter 7.

So, we can think about URIs as names for things, or names for relationships between things, rather than just as addresses for Web documents. Since every owner of a Web domain name can create its own unique URIs, this system is a simple way to create globally unique names across the whole of the Web—unlike Merea Maps'

Web service in our example, which could only supply identifiers that were local to its organization. When URIs are dereferenceable, they also act as a means of locating on the Web the resource that they identify.

Let us assign the URI http://mereamaps.gov.me/topo/0012 to the Isis Tavern pub from our example (our resource of interest on the Web). It is not exactly snappy, is it? Apart from taking up lots of space on the page, the longer the name is, the higher the likelihood of introducing typographical errors, and the more long-winded any file becomes that has to use it. To address this problem, a prefix is often used instead of the namespace. A namespace is the scope in which the name is unique, in this case http://mereamaps.gov.me/topo/. So, we could assign the prefix "mm" to be used instead of http://mereamaps.gov.me/topo/ and just write the Isis Tavern's URI as mm:0012.

Now that we have created a unique identifier for the Isis Tavern, we are ready to move on to model some information about it in RDF.

5.4   THE RDF DATA MODEL

RDF models data as a directed graph made up of nodes (the circles) and arcs (the links or graph "edges"), which have a direction: going out of one node and into the other node.

Figure 5.1 uses the URI we assigned to the Isis Tavern to represent its data as a network or graph. This graph can be broken down into a number of statements made up of a subject; a verb or verb phrase, known as a "predicate"; and an object. These three together are often known as a "triple." The graph of 5.1 can be written as a number of triples, with the pub's URI as the subject of each triple. The triples that describe the Isis Tavern pub are shown in Table 5.2.

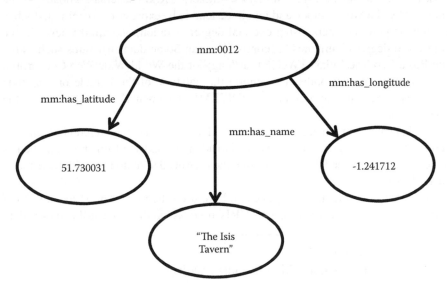

where mm: is the prefix representing http://mereamaps.gov.me/PlacesOfInterest/

FIGURE 5.1   The Isis Tavern represented as an RDF graph.

TABLE 5.2

RDF Triples Describing the Isis Tavern Pub

Subject	Predicate	Object
http://mereamaps.gov.me/topo/0012	http://mereamaps.gov.me/topo/has_name	"The Isis Tavern"
http://mereamaps.gov.me/topo/0012	http://mereamaps.gov.me/topo/has_latitude	51.730031
http://mereamaps.gov.me/topo/0012	http://mereamaps.gov.me/topo/has_longitude	−1.241712

Notice that the graph's nodes represent the subject or object of the triple, while the graph arcs are the predicates. A subject can be a URI that represents either a specific thing (known as an individual) such as a pub like the Isis Tavern or a class of things such as Pub. A predicate is also sometimes called a "property" or "relationship," and an object is also known as a "property value." URIs are used to name both resources and properties. An object can be a URI for a class or individual, or it can also be just a simple value (e.g., a string or number) that is known as a "literal," for example, the name "The Isis Tavern." The literal can be plain or typed. A plain literal is a string, which may have an optional language tag, indicating the language in which it is written, such as English or German. A typed literal is again a string, combined with a datatype URI to specify whether the string should be interpreted as an integer, a floating point number, a date, or some other type. The datatype URI can be any URI, but often a URI from the XML (eXtensible Markup Language) Schema standard is used. In our example, the latitude and longitude could be typed literals, expressed as floating point numbers, using "http://www.w3.org/TR/xmlschema-2/#float." At the moment, on the Web, latitude and longitude tend to be expressed as decimal, double, or floating point numbers, using decimal degree notation, and almost never in the traditional degrees ° minutes ′ seconds ″ format. Some data structures such as the the Basic Geo Vocabulary (WGS84 lat/long)[2] of the World Wide Web Consortium (W3C) simply do not bother to mention the datatype of the latitude or longitude to "reduce the syntactic burden on RDF/XML documents," that is, to make the RDF shorter.

As indicated, an object can take a URI as its value that references another resource. For example, if our user wanted to add in the extra information that the Isis Tavern was next to the River Isis, all that the user needed to do is to add the triple as in Table 5.3.

Note that this time, to save space, we have used the prefix "mm" for our http://mereamaps.gov.me/topo/namespace. URIs used as predicates can be prefixed in

TABLE 5.3

RDF Triple with a Resource Object

Subject	Predicate	Object
mm:0012	mm:is_next_to	http://sws.geonames.org/2636063

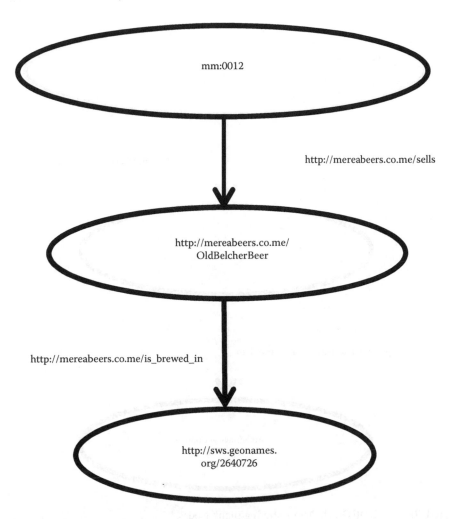

FIGURE 5.2 RDF graph showing "The Isis Tavern sells Old Belcher Beer brewed in Mereashire."

exactly the same way. In Table 5.3, the URI of the object identifies the River Isis from the well-known GeoNames RDF dataset. This is only possible in a relational database if the table structure itself is changed; in RDF, we just add more data.

An RDF statement can only model a relationship between two things, such as between the pub and the river. If we want to model how three or more things relate together, what is known as an "*n*-ary relation," for example to say, "The Isis Tavern sells Old Belcher Beer brewed in Mereashire," we need to introduce an intermediate resource. Let us pick this example apart a bit more: We have the first statement, "The Isis Tavern sells Old Belcher Beer," and a second one that says: "Old Belcher Beer is brewed in Mereashire." In an RDF graph, this is fairly straightforward, as shown in Figure 5.2, where Old Belcher Beer is represented as a class. So, the second statement says that every instance of Old Belcher Beer, that is, every pint and every barrel, is brewed in Mereashire.

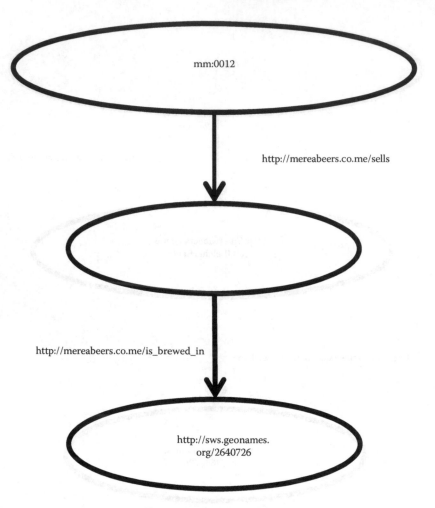

FIGURE 5.3   RDF Graph Demonstrating Blank Nodes.

But, what happens if we want to say something like, "The Isis Tavern sells beer that's brewed in Mereashire"? We know that it is some sort of beer, but we cannot just use the class Beer because it is certainly not true that all Beer is brewed in Mereashire. All we can say is that an instance of a beer is brewed there. However, since we do not know which one, we cannot give it an identifier. Instead, to model this, we introduce what is known as a "blank node" or "bnode" for short. A blank node is a node that acts as the subject or object of a triple but does not have its own URI—it is anonymous. In a graph, it is easy to represent: We just do not bother labeling the blank node, as in Figure 5.3.

But, if we write the same information in triples, we need to introduce a blank node identifier, which is of the form "_:name." For example, our beer blank node could be "_:beer0." This allows us to identify it locally within our dataset as there could be a number of blank nodes in the same graph. Again, the second triple in Table 5.4 is calling on the GeoNames dataset to supply a URI for "Mereashire."[3]

TABLE 5.4

RDF Triples Denoting the Information "The Isis Tavern sells beer that's brewed in Mereashire"

Subject	Predicate	Object
mm:0012	http://mereabeers.co.me/sells	_:beer0
_:beer0	http://mereabeers.co.me/is_brewed_in	http://sws.geonames.org/2640726

This is only one type of n-ary relation; there are others that can only be modeled in OWL (Web Ontology Language) DL and not in RDF alone. These are discussed further in Chapter 10. Note a couple of final points about blank nodes before we move on: First, a blank node is not accessible outside the graph and hence cannot be considered during data integration. Querying graphs containing a lot of blank nodes is slower. Second, blank nodes can only be used for subjects or objects; we cannot have blank properties.

5.5 RDF SERIALIZATION

RDF on its own is a data *model*, not a data format. Serialization is the process of putting an RDF graph into a data format to publish it on the Web. That is, you take the triples and put them in a file, using a particular syntax. This can be done in advance, off line, if the dataset is static, or it can be done on demand if the data set is known to change a lot. As usual, there are plenty of different RDF formats. W3C has standardized two syntaxes (RDF/XML and RDFa [Resource Description Framework–in–attributes]), and there are other nonstandard serialization formats that offer advantages in certain circumstances.

5.5.1 RDF/XML

RDF/XML is the "normative" or typical syntax that is used to format RDF (Beckett, 2004). W3C has specified a set of XML tags, or "vocabulary," that can be used to describe the elements in an RDF graph, which are denoted by the RDF namespace http://www.w3.org/1999/02/22-rdf-syntax-ns#. It is not a great idea, although you can just open up a text editor and start typing in all your data as RDF/XML. It is verbose and long winded, and you will end up making mistakes. As discussed further in Chapter 7, there are plenty of RDF editors with graphical user interfaces that will output RDF/XML without you having to author it directly, such as Altova's SemanticWorks, Topbraid Composer, Rej, or Protégé (of which there are two versions, offering different features), and there are also tools to convert your data sitting in a GIS (Geographic Information System) or relational database into RDF. So, the aim of this section is to help you understand the RDF/XML syntax when you see it, rather than expecting you to jump right in there and author the RDF/XML directly. Also, here is a word of warning: If you are coming from a background in XML, you should probably avoid using RDF/XML and choose one of the alternative formats, such as Turtle (Terse RDF Triple Language; see Section 5.5.2), for authoring and data management instead. This will help you concentrate on the *knowledge you are encoding in the graph structure* rather than the details of the XML markup tags.

To explain what the different tags are used for, let us look at an example:

```
1   <?xml version = "1.0" encoding = "UTF-8"?>
2   <rdf:RDF
3      xmlns:rdf = "http://www.w3.org/1999/02/22-rdf-syntax-ns#"
4      xmlns:mereaMaps = "http://mereamaps.gov.me/topo/">
5     <rdf:Description
6        rdf:about = "http://mereamaps.gov.me/topo/0012">
7        <rdf:type rdf:resource = "http://mereamaps.gov.me/topo/Pub"/>
8        <mereaMaps:has_name>The Isis Tavern</mereaMaps:has_name>
9        <mereaMaps:has_longitude>-1.241712</mereaMaps:has_longitude>
10    </rdf:Description>
11  </rdf:RDF>
```

Line 1 is just standard XML, saying that the file is in XML format and specifying the XML version. The tag at line 2 `rdf:RDF` (which is closed at line 11 using the standard XML convention of backslash followed by the same tag `</rdf:RDF>`) tells us that this XML document represents an RDF model. In general, the `<rdf:RDF>` tag should always be the first or "root" element of your RDF/XML document. On lines 3 and 4, we specify prefixes for two namespaces: the rdf: prefix, which is used to represent the RDF namespace, and our example prefix mereaMaps: for Merea Maps' Topo namespace. From now on in the document, we know that any term that starts with rdf: is part of the RDF vocabulary (or tag set), and any term that starts with mereaMaps: is part of the mereaMaps namespace. Line 5 uses a tag from the RDF vocabulary: `rdf:Description`; everything between this and its closing tag on line 10 is part of this RDF Description. Line 6 `rdf:about` tells us which resource the `rdf:Description` is talking about, in this case, the URI of the Isis Tavern: http://mereamaps.gov.me/topo/0012.

Line 7 shows how RDF enables us to give classes their own URIs so that we can say that the Isis Tavern *is a* Pub. (You should read the predicate "is a" as "is an instance of.") Further in this chapter we talk about how RDFS allows us to specify a number of these classes, and build up hierarchies, to say for example, that a Pub *is a kind of* Topographic Object (where the predicate "is a kind of" is used to denote "is a member of the class"). But for now, we can just note that the RDF tag `rdf:type` lets us specify that the Isis Tavern is of the type (class) Pub.

Finally, lines 8 and 9 specify properties and property values (i.e., predicates and objects from the original RDF triples) to state the name and longitude of the Isis Tavern. The URI of the triple's subject—the Isis Tavern—does not need to be repeated for each statement since it is all enclosed within the `rdf:Description` tag.

So, this small amount of code is our first RDF/XML document. We now build up a more complete picture of the RDF/XML syntax by adding in some extra statements. First, let us go back to the following example in which a triple has a resource URI as an object. To indicate that the Isis Tavern is next to the River Isis, in RDF/XML we can say

```
1   <?xml version = "1.0" encoding = "UTF-8"?>
2   <rdf:RDF
3       xmlns:rdf = "http://www.w3.org/1999/02/22-rdf-syntax-ns#"
4       xmlns:mereaMaps = "http://mereamaps.gov.me/topo/">
5     <rdf:Description
6       rdf:about = "http://mereamaps.gov.me/topo/0012">
7       <mereaMaps:is_next_to
8         rdf:resource = "http://sws.geonames.org/2636063"/>
9     </rdf:Description>
10  </rdf:RDF>
```

So, lines 7 and 8 use the `rdf:resource` tag with the GeoNames URI for the River Isis. Next, let us look at how to express the graph from Figure 5.3 containing a blank node in RDF/XML syntax, using the RDF tag `rdf:nodeID`, as follows:

```
1   <?xml version = "1.0" encoding = "UTF-8"?>
2   <rdf:RDF
3       xmlns:rdf = "http://www.w3.org/1999/02/22-rdf-syntax-ns#"
4       xmlns:mereaMaps = "http://mereamaps.gov.me/topo/">
5     <rdf:Description
6       rdf:about = "http://mereamaps.gov.me/topo/0012">
7       <mereaMaps:sells
8         rdf:nodeID = "beer0"/>
9     </rdf:Description>
10    <rdf:Description rdf:nodeID = "beer0">
11      <mereaMaps:is_brewed_in
12        rdf:resource = "http://sws.geonames.org/2640726"/>
13    </rdf:Description>
14  </rdf:RDF>
```

On line 8 the `rdf:nodeID` is used to denote the blank node with local name "beer0." (The predicate `rdf:nodeID` is used for blank nodes, while `rdf:ID` can be used for named nodes, where the URI supplied is an absolute one, that is, is not relative to any initially supplied namespace prefix. rdf:about can be used for naming nodes with a URI relative to a namespace prefix.) Then, in lines 10 to 13, the blank node "beer0" gets its own `rdf:Description` in order to state that it is brewed in the GeoNames resource of Mereashire. Note that this is an example of how you can include a description of multiple resources in the same RDF/XML document by just listing their `rdf:Descriptions` one after another.

Other terms in the RDF/XML vocabulary that you may come across include `rdf:datatype`, `rdf:parseType`, and `rdf:value`, which can be used to encode a literal object that has a particular datatype (such as string, integer, or floating point number). The following is an example of this, to state that the Isis Tavern's latitude is expressed as a floating point number, of unit degrees, and the Isis Tavern's name is a string:

```
1   <?xml version = "1.0" encoding = "UTF-8"?>
2   <!DOCTYPE
3   <rdf:RDF [!ENTITY xsd http://www.w3.org/2001/XMLSchema#>]>
4   <rdf:RDF
5       xmlns:rdf = "http://www.w3.org/1999/02/22-rdf-syntax-ns#"
6       xmlns:mereaMaps = "http://mereamaps.gov.me/topo/">
7     <rdf:Description
8       rdf:about = "http://mereamaps.gov.me/topo/0012">
9       <mereaMaps:has_name
10        rdf:datatype = "&xsd;string"/>The Isis Tavern
11      </mereaMaps:has_name>
12      <mereaMaps:has_latitude rdf:parseType = "Resource">
13        <rdf:value
14          rdf:datatype = "&xsd;float"> 51.730031</rdf:value>
15        <http://mereamaps.gov.me/has_units
16          rdf:resource = "http://mereamaps.gov.me/units/degrees"/>
17      </mereaMaps:has_latitude>
18    </rdf:Description>
19  </rdf:RDF>
```

Lines 2 and 3 specify that the reference name xsd can be used in place of the XML entity http://www.w3.org/2001/#, which is just a way of abbreviating the datatypes http://www.w3.org/2001/#string and http://www.w3.org/2001/#float in lines 10 and 14. Lines 9 to 11 state that the Isis Tavern has the name "The Isis Tavern," which is a string datatype. Lines 12 to 17 state, through the use of a blank node that *is not* assigned a local identifier, that the Isis Tavern has a latitude of 51.730031 that is a floating point datatype and has units (as specified by Merea Maps' units namespace) of degrees.The last piece of RDF/XML syntax that we discuss here concerns *reification*. Reification is the process of making statements about statements. This is particularly useful for adding provenance information about the RDF triple, such as who wrote it and when. A word of warning though: Reification causes problems when querying, so it is not recommended for use in Linked Data. In Chapter 7, we discuss alternatives for including provenance information without the use of reification. For now, though, it is worth being able to recognize it. The following XML fragment gives an example of reification, showing that the person codeMonkey1 at Merea Maps created the triple: mereaMaps:0012 mereaMaps:has _ name "The Isis Tavern." A predicate dc:creator is used, which comes from the Dublin Core vocabulary discussed in a further part of this chapter.

```
1   <?xml version = "1.0" encoding = "UTF-8"?>
2   <rdf:RDF
3       xmlns:rdf = "http://www.w3.org/1999/02/22-rdf-syntax-ns#"
4       xmlns:mereaMaps = "http://mereamaps.gov.me/topo/"
5       xmlns:dc = "http://purl.org/dc/elements/1.1/">
```

```
6    <rdf:Statement>
7      <rdf:subject rdf:resource = "mereaMaps:0012"/>
8      <rdf:predicate rdf:resource = "mereaMaps:has_name"/>
9      <rdf:object rdf:resource = "The Isis Tavern"/>
10     <dc:creator rdf:resource = "http://mereamaps.gov.me/
codeMonkey1"/>
11   </rdf:Statement>
12 </rdf:RDF>
```

RDF/XML is commonly used to publish Linked Data, but its syntax is quite verbose, making it difficult for people to read and write. It is worth considering an alternative serialization for managing and curating RDF data, where someone has to interact with the data, or for presenting it to the end user, where the end user needs to be able to understand it easily. We talk more in Chapter 7 about how to author and validate your own RDF data, but for now, let us look at the other RDF formats.

5.5.2 TURTLE

Turtle is the Terse RDF Triple Language (Beckett and Berners-Lee, 2008), which is in the process of being standardized by W3C and is widely used for reading and writing triples by hand as it is shorter and in plaintext. (If you feel a sudden urge to get out your text editor and start writing RDF, then this is the format to use.) The RDF query language SPARQL, which is discussed in Chapter 7, borrows a lot of its syntax from Turtle. An example of Turtle is as follows:

```
1  @prefix rdf: <http://www.w3.org/1999/02/22-rdf-syntax-ns#>.
2  @prefix mereaMaps: <http://mereamaps.gov.me/topo/>.
3
4  mereaMaps:0012 mereaMaps:has_name "The Isis Tavern".
5  mereaMaps:0012 mereaMaps:has_longitude "-1.241712".
```

Notice the prefixes, which avoid the need to repeat long namespaces. URIs are always in angle brackets when written in full, as in lines 1 and 2, and statements end with a full stop, as in line 4 and 5. Turtle's advantage lies in its shortcuts, making it quicker to read and write. These include the use of a default namespace, which does not require an explicit prefix but just uses a colon as follows:

```
1  @prefix : <http://mereamaps.gov.me/topo/>.
2  @prefix geonames: <http://sws.geonames.org/>.
3
4  :0012 a :Pub;
5      :has_name "The Isis Tavern"@en;
6      :has_longitude "-1.241712"^^<http://www.w3.org/2001/
XMLSchema#float>;
7      :sells [
8         :is_brewed_in geonames:2640726.
9      ].
```

In line 1, we make http://mereamaps.gov.me/topo the default namespace. Multiple triples with the same URI subject can be grouped into one statement, delineated by a semicolon if they have different predicates and objects, as in lines 4, 5, and 6, or by a comma if they have the same predicate but just different objects. The predicate "a" is used instead of `rdf:type`, as in line 4, to show that the Isis Tavern is a Pub. The @ sign is used to indicate language, as in line 5, which states that "The Isis Tavern" is in English, and ^^ is used to indicate a datatype, as in line 6. Square brackets are used for blank nodes, as in lines 7, 8, and 9. The file extension for turtle files is.ttl.

5.5.3 N-Triples

N-Triples (Grant and Beckett, 2004) is a subset of Turtle (and not to be confused with the Notation 3 language, which is a superset of Turtle, also incorporating support for rules (Berners-Lee, 1998a). We have left Notation 3 outside the scope of this book as the full language is less widely used or supported than standard RDF). N-Triples files should use the .nt file extension to distinguish them from Notation 3, which uses the .n3 suffix. The N-Triples syntax does not allow prefixes or other shortcuts but requires every triple to be written out in full, as in

```
1   <http://mereamaps.gov.me/topo/0012> <http://www.w3.org/1999/
02/22-rdf-syntax-ns#type> <http://mereamaps.gov.me/topo/Pub>.
2   <http://mereamaps.gov.me/topo/0012> <http://mereamaps.gov.me/ topo/
has_name> "The Isis Tavern".
3   <http://mereamaps.gov.me/topo/0012> <http://mereamaps.gov.me/ topo/
has_longitude> "1.241712" ^^<http://www.w3.org/2001/XMLSchema#float>.
4   <http://mereamaps.gov.me/topo/0012> <http://mereamaps.gov.me/ topo/
sells> _:beer0.
5   _:beer0 <http://mereamaps.gov.me/topo/isbrewed_in>
<http://sws.geonames.org/2640726>.
```

which contains the same triples as the last example. This makes an N-Triples file quite long but means that each line in the file can be parsed separately. This means that the N-Triples format is particularly good for large data files that cannot be loaded into memory all at once, for parallel processing of the data or for backing it up.

There are tools on the Web that automatically convert from Turtle and N3 to RDF/XML format and vice versa,[4] making it easy to write Turtle and yet publish in RDF/XML. There is also a push from the W3C RDF Working Group to standardize a JavaScript Object Notation (JSON) serialization for RDF, which would mean that Web developers programming in JavaScript, Ruby, or PHP (hypertext processor) could use RDF without needing additional software libraries.

5.5.4 RDFa

As mentioned in Chapter 2, RDF can also be embedded directly in HTML documents, rather than maintaining a separate file, using RDFa, which is the other W3C standard format for RDF (Adida and Birbeck, 2008). This format allows existing

Web pages to be marked up with RDF, making it easier to include structured data into preexisting Web content publishing workflows.

```
1   <!DOCTYPE html PUBLIC "-//W3C//DTD XHTML+RDFa 1.0//EN"
2     "http://www.w3.org/MarkUp/DTD/xhtml-rdfa-1.dtd">
3   <html xmlns = "http://www.w3.org/1999/xhtml"
4       xmlns:rdf = "http://www.w3.org/1999/02/22-rdf-syntax-ns#"
5       xmlns:mereaMaps = "http://Mereamaps.gov.me/topo/"
6       xmlns:dc = "http://purl.org/dc/elements/1.1/">
7
8   <head>
9   <meta http-equiv = "Content-Type" content = "application/xhtml+xml;
10    charset = UTF-8"/>
11      <meta property = "dc:creator" content = "CodeMonkey1"/>
12      <title>The Isis Tavern Homepage </title>
13  </head>
14
15  <body>
16    <div about = "mereaMaps:0012" typeof = "mereaMaps:Pub">
17      <span property = "mereaMaps:has_name" content = "The Isis
Tavern"/>
18      <span property = "mereaMaps:has_longitude" content =
"-1.241712"/>
19      <span property = "mereaMaps:has_latitude" content = "51.730031"/>
20      <h1> Welcome to the Isis Tavern!</h1>
21    </div>
22  </body>
23  </html>
```

This is an example of an HTML document containing RDFa. This could be a Web page for the Isis Tavern, which contains the following embedded information:

a. CodeMonkey1 created this page (line 11).
b. There is content about a Merea Maps resource 0012, which is a Pub (line 16).
c. Merea Maps resource 0012 has the name "The Isis Tavern" (line 17).
d. Merea Maps resource 0012 has the longitude –1.241712 and latitude 51.730031 (lines 18 and 19, respectively).

W3C's markup validation service[5] can check that your RDFa is correctly formatted. Just as few people these days author HTML directly, it is unlikely that you would author RDFa by hand. There are a few Web editors that output HTML+RDFa, such as Topbraid Composer, an XHTML+RDFa parent theme for Wordpress 2.7,[6] and an RDFa documents extension for Dreamweaver.[7]

5.6 RDFS

This section addresses how to model RDF data using the RDF Schema language RDFS. In previous chapters, we looked at data modeling from a Geographic Information (GI) perspective, but now we take a more computer science-based approach.

5.6.1 CONCEPTS AND INSTANCES: INSTANTIATION AND HIERARCHY IN RDFS

In our examples in this chapter, we have been working with data representing a real-world object, the Isis Tavern. The data was originally stored in a relational database, in a table containing data about Topographic Objects. We showed how to specify using RDF that the Isis Tavern was a Pub, and that Pubs were Topographic Objects. This introduces two ideas that it is vital to handle when dealing with the Semantic Web: *instantiation* and *hierarchy*. Instantiation simply means that a thing is a member of a category. In this book, we tend to refer to these categories as *Concepts,* when we are dealing with the GI world, or *Classes,* when taking a computer science approach. So, we have the class of Pubs, and we say that the Isis Tavern is an *instance* of a Pub; that is, it is a member of the class of Pubs.[8] This idea of instantiation uses the relationship "is a," which is short for "is an instance of." The second idea, hierarchy, may be more familiar to the reader and refers to when one class is a subclass of another; that is, it takes all the meaning of the superclass but also has additional, more specialized, properties. For example, Pub is a subclass of the Topographic Object class. We usually use the relationship "is a kind of" to describe this hierarchical relationship: "A Pub is a kind of Topographic Object." This means that the Pub class inherits all the statements made about Topographic Object and can have additional descriptions. Note that we have to be careful about statements that are made about a particular pub, for example, that the Isis Tavern is next to the River Isis, versus statements that hold true for *all* pubs, for example, that Pubs sell Beer. If we state that all Pubs sell Beer, and that the Isis Tavern is a Pub, then we can infer that the Isis Tavern sells Beer. However, saying that the Isis Tavern is next to the River Isis, and that the Isis Tavern is a Pub, only tells us that *at least one* Pub (the Isis Tavern in this case) sells Beer, not necessarily that *all* of them do. We talk more about this sort of logical inference in Chapter 9 regarding the topic of OWL.

5.6.2 VOCABULARIES AND ONTOLOGY

RDF allows us to make statements, in the form of triples, about things in our *domain,* or subject of interest; but to specify which classes we expect in the domain and what properties can be used with each class, we need something extra: an ontology.

While a *vocabulary* is the set of terms (class and property names) that can be used in the domain, an *ontology*[9] spells out how these vocabulary terms can be used with each other—it specifies the permitted structure. For example, a vocabulary for the topographic domain would include the terms "TopographicObject," "Pub," "has _ longitude," "has _ latitude," "has _ name," and so on. An ontology would say that Pub and TopographicObject are classes, that Pub is a subclass of TopographicObject, and that the class TopographicObject can have the properties "has _ longitude," "has _ latitude," and "has _ name."

We can state all these things using the RDFS language, that is, which classes we have and which properties each class is allowed to take. The rules of inheritance mean that a property of a class will also be a property of its subclass. This makes sense when thinking about an example: If TopographicObject has the property "has _ longitude," and a Pub is a kind of TopographicObject, then Pub can also use the property "has _ longitude." A second rule states that if, for example, Pub is a kind of TopographicObject, and Freehouse[10] is a kind of Pub, then Freehouse is also a kind of TopographicObject. This kind of inference is known as *transitivity*, so we can say that the rdfs:subClassOf relationship is transitive.

5.6.3   RDFS Syntax: Classes and Properties

Let us use an example to introduce the syntax of RDFS.

```
1   <?xml version = "1.0" encoding = "UTF-8"?>
2   <rdf:RDF
3       xmlns:rdf = "http://www.w3.org/1999/02/22-rdf-syntax-ns#"
4       xmlns:rdfs = "http://www.w3.org/200/01/rdf-schema#"
5       xmlns:mereaMaps = "http://mereamaps.gov.me/topo/">
6
7   <rdf:Description
6       rdf:about = "http://mereamaps.gov.me/topo/Pub">
8       <rdf:type
9          rdf:resource = "http://w3.org/2001/01/rdf-schema#Class"/>
10       <rdfs:subClassOf rdf:resource = "mereaMaps:TopographicObject"/>
11   </rdf:Description>
12  </rdf:RDF>
```

The language constructs in RDFS are themselves classes and properties. Any application that knows the vocabulary will know what classes and properties to expect and can process the data, including from multiple documents using the same ontology. The two basic classes within the RDFS language are

- rdfs:Class, which is the class of resources that are RDF classes
- rdf:Property, which is the class of all RDF properties

The example states that Pub is a Class (using the rdf:type property and rdfs:Class as the object of the triple, lines 8 and 9) and a subclass of Topographic Object (using the rdfs:subClassOf property, line 10).

A new property is defined by assigning it a URI and describing it with an
`rdf:type` property whose value is the resource `rdf:Property`. For example,
we can say that "has _ name" is a kind of `Property` in the following way:

```
<rdf:Property rdf:ID = "has_name">
   <rdfs:domain rdf:resource = "mereaMaps:TopographicObject"/>
</rdf:Property>
```

5.6.4 SUBPROPERTIES, DOMAIN, AND RANGE

RDFS also includes several properties, for example,

- rdfs:subPropertyOf
- rdfs:range
- rdfs:domain

Just as classes can have subclasses, so properties can have subproperties, denoted
by `rdfs:subPropertyOf`. For example, has _ preferred _ name and
has _ short _ name might be subproperties of has _ name. One advantage
of this is that we can differentiate between the different types of name but still
retrieve them all using the superproperty has _ name. If we want to state that the
subject of a particular property is always an instance of a particular class, we use
the `rdfs:range` property. Similarly, the `rdfs:domain` property allows us to
specify what class the object of the triple has to be.

```
<rdf:Property rdf:ID = "has_longitude">
   <rdfs:domain rdf:resource = "mereaMaps:TopographicObject"/>
   <rdfs:range rdf:resource = "&xsd;float"/>
</rdf:Property>
```

Here, we are saying that any instance that is the object of the "has _ longitude"
property is of floating point type, and any instance that is the subject of the
"has _ longitude" property must be a `TopographicObject`. Note that this
is a far stricter statement than just saying that the has _ longitude property
can be used by a `TopographicObject`, and as you can see for this example, it
is not really true: So often when authoring your RDFS ontology and modeling the
knowledge, it is best not to use domain and range unless you are absolutely sure that
these things are true and true for every instance of the class. We talk more about
knowledge modeling when we write ontologies in Chapter 10.

5.6.5 RDF CONTAINERS AND COLLECTIONS

It is possible to denote groups of resources or values ("literals") using one of the
RDF container constructs `rdf:Bag` (a group of resources or literals for which the
order does not matter, and duplicates are allowed); `rdf:Seq` (an ordered sequence

of resources or literals for which duplicates are allowed); and `rdf:Alt` (a group that represents alternatives to each other that does not allow duplicates). While an RDF container only specifies that the named resources are members of that group, it does not preclude other resources, not mentioned, from also being members of the group. An RDF collection, that is, a closed, ordered group of resources or literals, can be used to specify "this is every single member in the group," using the construct `rdf:List`. Note that RDF containers and collections are hard to query with SPARQL, so are best avoided if the data is to be published as Linked Data.

5.6.6   RDFS Utility Properties

The last set of RDFS constructs that you will need to be aware of are called "utility" properties as they help to add metadata to the ontology. They are:

- `rdfs:seeAlso`. A property that can be used on any resource to say that another resource *may* provide additional information about that thing. For example, the following is a triple expressed in Turtle syntax that suggests that the mypubreviews dataset will provide additional information about the Isis Tavern.

```
<http://mereamaps.gov.me/topo/0012> rdfs:seeAlso
<http://mypubreviews.com/isisTavern>.
```

- `rdfs:isDefinedBy`. This is a subproperty of `rdfs:seeAlso` and is used to indicate that the object of the triple is the original and authoritative description of the resource. So, the mypubreviews site may wish to state that Merea Maps is the authoritative source of information about the Isis Tavern:

```
<http://mypubreviews.com/isisTavern>
   rdfs:isDefinedBy <http://mereamaps.gov.me/topo/0012>.
```

- `rdfs:comment` provides a human-readable description of the property or class in question. For example,

```
<http://mereamaps.gov.me/topo/> rdfs:comment "An ontology describing
the topographic geography in Merea."@en.
```

- `rdfs:label` is used to provide a human-friendly name for the class. For example, while the class might, in computer speak, be called `TopographicObject`, the `rdfs:label` could name it using the string "Topographic Object."

Using the RDFS constructs, a concept can be described as a combination of its class, what that is a subclass of, which properties it can have, and what values those properties can take. It is the set of these descriptions for concepts in a domain that make up the ontology. We look in more detail at ontologies in Chapters 9 and 10.

5.7   POPULAR RDFS VOCABULARIES[11]

In the topographic RDFS ontology example, we have coined (or, to use the Linked Data term, we have "minted") our own URI for the "has _ name" property, which we can almost certainly expect to have been defined by someone else previously. It is best practice to reuse common properties like "has _ name" rather author our own, provided that we agree with the original author's understanding of that property. It is less work and means that our data integrates better with other data published on the Linked Data Web. In Chapter 8, we talk about how to go about finding out whether someone else has already authored a URI for a property you want to use and how to double-check that they are using it to mean exactly the same as you. For now, we review some basic RDFS vocabularies that contain terms like "name" that you may well wish to reuse.

Dublin Core is an International Organization for Standardization (ISO) standard (ISO 15836:2009) set of fifteen metadata elements such as title, creator, date, and rights. The RDF version of Dublin Core is at http://purl.org/dc/elements/1.1/. One term of particular interest is "coverage," which refers to the "spatial or temporal topic of the resource, the spatial applicability of the resource, or the jurisdiction under which the resource is relevant." This can be useful in specifying the context in which your ontology can be used.

SKOS (Simple Knowledge Organization System) (Miles and Bechhofer, 2009) was designed to provide a simple way to port existing knowledge representation systems (primarily thesauri and taxonomies), like those used in libraries, social tagging, or other simple classification systems, to the Semantic Web. It is a W3C specification and includes properties such as skos:broader and skos:narrower to describe the relationship between one genre and its more specific species or between a whole and its parts. Whereas RDF would distinguish between hierarchical (subclass of) and mereological (part of) relationships, the SKOS terms are much less precise, and hence SKOS may be more suitable for use where the exact relationship is not known. Similarly, where RDF might coin a term "participates in" to indicate the relationship between an event and a group of entities that typically participate in it, SKOS uses the vaguer skos:related property, which can be used to indicate any sort of association between two concepts. SKOS also offers some interesting properties to be used with metadata. For example, skos:scopeNote indicates the limits of where the concept can be used; skos:definition is used in the same way as rdfs:comment (to store a human-readable text description of the concept); and skos:example gives an example of how the concept should be used.

To map between different ontologies, SKOS uses skos:exactMatch and skos:closeMatch. skos:exactMatch is used to link classes that are equivalent in meaning, for example, Car and Automobile, but unlike owl:sameAs, which we discuss in Chapter 8, it is much vaguer and does not imply anything about the two concepts sharing all their statements, as is required by the logic of OWL. skos:closeMatch is even vaguer and just indicates that the two concepts are similar in meaning. skos:broadMatch, skos:narrowMatch, and skos:relatedMatch are also available. A new ontology can reuse existing concepts using the skos:inScheme property.

For our purposes in GI, SKOS is of most interest when we want to indicate some relationship between vocabularies or ontologies where we do not have enough information to specify exact links or where the vocabularies themselves have come from imprecise data, such as crowd-sourced tags.

FOAF, the Friend of a Friend project (Brickley and Miller, 2010), defines an RDF vocabulary for expressing metadata about people and their interests, relationships, and activities. The main class that is used is foaf:Person, which can take properties like foaf:name, foaf:gender, foaf:member, foaf:organization, and foaf:mbox, the last indicating the person's e-mail address. We might reuse foaf:name in our Topo ontology instead of defining our own "has _ name" property, for example.[12] FOAF does not assign a URI to a person, however, because there is still a debate about assigning one URI to a person. Who should mint it? What happens when a person has multiple URIs? To get around this problem, FOAF places a restriction on certain properties[13] so that when a property is used in a triple, the property's object value uniquely defines the subject resource. One example of this is foaf:mbox (the property "has email address"), implying that an e-mail address can be said to pinpoint one specific person. If one person called Bob has an e-mail address bob@example.com, and another person called Robert has an e-mail address bob@example.com, then we can infer that Bob and Robert are the same person. This is also an example of where FOAF may not have gotten things right as there are many occasions when people share e-mail addresses; therefore, the inference would be incorrect.

The other frequently used property that FOAF provides is foaf:knows, which specifies that person A knows person B (although it may be unrequited; person B may deny all knowledge of person A in return). This can then be used to build a graph of the social networks between people, which can be useful in many applications.

5.7.1 Geo RDF

The W3C's Basic Geo WGS84 lat/long vocabulary[14] is widely used on the Linked Data Web to represent very basic spatial location information, using WGS84 as a reference. It only includes the classes geo:SpatialThing and geo:Point. A geo:Point can have the relationships geo:lat (for latitude), geo:long (for longitude), and geo:alt (for altitude). Our Pub example could hence be rerendered using the "geo" and "foaf" namespaces as

```
1   <?xml version = "1.0" encoding = "UTF-8"?>
2   <rdf:RDF
3      xmlns:rdf = "http://www.w3.org/1999/02/22-rdf-syntax-ns#"
4      xmlns:geo = "http://www.w3.org/2003/01/geo/wgs84_pos#"
5      xmlns:foaf = "http://xmlns.com/foaf/0.1/">
6   <rdf:Description
7      rdf:about = "http://mereamaps.gov.me/topo/0012">
8      <rdf:type rdf:resource = "http://mereamaps.gov.me/topo/Pub"/>
9      <foaf:name>The Isis Tavern</foaf:name>
10     <geo:long>-1.241712</geo:long>
```

```
11      <geo:lat>51.730031</geo:lat>
12    </rdf:Description>
13  </rdf:RDF>
```

The GeoOnion RDF/XML vocabulary,[15] which takes the namespace "go:," was an early attempt, as yet incomplete, to provide a number of properties that relate spatial things together based on their distance in meters. It suffers from a lack of GI input, as the initial idea was to specify nearness in terms of concentric circles around the point of interest. This is a woefully inadequate way to deal with the richness of the nearness relationship. It is in exactly this kind of initiative that the GI specialist's experience of the contextual dependence of geographical nearness could come into play in modeling knowledge more accurately.

In addition, there are several OWL ontologies that deal with spatial information, such as the geoFeatures,[16] geoCoordinateSystems,[17] and geoRelations[18] ontologies.

5.8 RDF FOR THE THINKING GEOGRAPHER

So, from the geographer's point of view, what are the most important differences between RDF and traditional methods of representing data in the geographical field? As we have seen, the RDF data model is a graph, which makes it far more amenable to the addition of new information than the more familiar tabular form used to represent data in a GIS sitting on top of a relational database. While adding a new property to a geographical feature (that is, adding a new column to a relational database table) means that every geographical feature must provide a value for that field, or include a null, thus increasing the size of the table significantly, this is not the case in RDF, where only new triples need be added to the graph, without the need to pad a table.

Identity is another topic that the geographer must treat carefully. A real-world object may well be given an identifier in a GIS, and there have been efforts made to create a global system of identifiers. However, RDF has a ready-made, and superior, system for minting identifiers through URIs. In many cases, an organization can use its own internal system for identifier creation, plus their Web domain name, to create public URIs for their data. The process of discerning whether two things are the same is an outstanding problem both for the Semantic Web and for a GIS specialist. While in GIS spatial collocation is often used as the primary factor in determining an equivalence relationship, this is far less likely to be a sufficient indicator in the semantic world. For example, a building in a dataset mapped 100 years ago with exactly the same spatial footprint as the building today might well be identified as the same building by a GIS even though its usage had changed. However, RDF might have two URIs, one for a Prison, which the building was used as last century, and one for a Hotel, which is the building's current use. RDF could also have a third URI that represents the building irrespective of its use. RDF places far more emphasis on *context*, whereas a GIS is mainly concerned with spatial information.

As we explain in Chapter 9, using the owl:sameAs relationship between two URIs is quite a strong assertion to make as *all* the statements about one URI are then known to be true about the equivalent URI. However, room prices at the hotel are unlikely to be valid for the prison. It might be that we choose to model the knowledge as a single Building, with a current and a former use, but what if the building has been extended to make it into a hotel? Then, the spatial footprint will be different. In one context, the building is still the same but now has a different use and a different spatial footprint, but in another context, it can be thought of as two separate things. There is no easy answer to this knowledge modeling problem—we just point it out here to warn the reader that spatial collocation is not sufficient to establish equivalence, and while RDF offers greater opportunities for modeling the nuances of our knowledge, it also can create deeper pitfalls when equivalence relations are applied without due care and attention.

A GIS specialist is likely to be familiar with hierarchical relationships, equivalence or mereological ("part of") relationships; however, it is important when modeling data using RDFS to consider using other relationships that may well express your knowledge more accurately and not feel restricted to just hierarchical or mereological ones. For example, a Post Office could be "within" a Shopping Mall if it is located spatially inside the footprint of the Shopping Mall, rather than necessarily be considered "part of" it, and the two should certainly not be considered to be related in an explicit hierarchy. Since hierarchical relationships result in the child inheriting all the parent's properties, the "is a kind of" relationship of RDFS needs to be used with care. Remember that *any* arbitrary relationship can be created in RDF; hierarchy is not the only form of relationship!

RDF can offer some advantages to the geographer in modeling vagueness. For example, if a boundary cannot be precisely measured, a semantic term like "nearby" may be more appropriate for use as an RDF relationship than attempting to quantify vague boundaries explicitly and numerically. Alternatively, a vocabulary like SKOS could be used to indicate an unspecified similarity between two concepts.

5.9 SUMMARY

This chapter has explained how to model data in RDF using statements of triples and covered the basics of the RDF language, looking at the main varieties you might encounter: RDF/XML, Turtle, N Triples, and RDFa. We have also talked about how to use RDFS to specify which classes and properties you intend to use to structure your data in RDF. We covered some basic RDF vocabularies like Dublin Core, FOAF, and Geo and pointed out some potential traps to avoid as a geographer coming new to RDF data modeling. In the next chapter, armed with this grounding of RDF and basic knowledge representation in the triple structure, we explain how to organize GI as Linked Data.

NOTES

1. RDFS is a simple ontology language.
2. http://www.w3.org/2003/01/geo/#vocabulary
3. Before you go looking for Mereashire in GeoNames, please note this is a fictional example.
4. For example, http://rdfabout.com/demo/validator/.
5. http://validator.w3.org/.

6. http://www.sampablokuper.com/wp-content/uploads/2009/01/spk_xhtml_rdfa_1_
   parent.zip
7. http://www.adobe.com/cfusion/exchange/index.cfm?event=extensionDetail&loc=en_us
   &extid=1759526
8. As a convention, in this book we capitalize any words we are using as the name of a class.
9. You may also come across the term *taxonomy*; this denotes a controlled vocabulary that
   has been structured only into a hierarchy. Whereas with a fully fledged ontology any
   relationship can be used, a taxonomy is limited to the "is a kind of" relationship only.
10. In the United Kingdom, this is the term used for a pub that is not tied to one particu-
    lar brewery.
11. Very simple RDFS ontologies are often referred to as vocabularies, although in truth
    they are strictly more than just vocabularies. Nonetheless, this is what we will refer to
    them as in this section as this is how they are commonly known.
12. While it is common practice to exclude the auxiliary verb "has" or "is" from property
    names, we prefer to include it (e.g., "has_name," "is_located_in") as it enables automatic
    translation from a controlled natural language, and it reads better, making the relationship
    a bit more explicit, particularly for domain experts not well versed in ontology writing.
13. This type of restriction is known as an inverse functional property and is discussed further
    in Chapter 9.
14. http://www.w3.org/2003/01/geo/.
15. http://www.w3.org/wiki/GeoOnion
16. http://www.mindswap.org/2003/owl/geo/geoFeatures.owl
17. http://www.mindswap.org/2003/owl/geo/geoCoordinateSystems.owl
18. http://www.mindswap.org/2003/owl/geo/geoRelations.owl

6 Organizing GI as Linked Data

6.1 INTRODUCTION

This chapter provides advice on how to represent Geographic Information (GI) as Linked Data. We cannot hope to cover all aspects of this representation; the topic is simply too broad for any one book and individual experience. Therefore, we concentrate on some of the essential areas. The chapter explicitly examines identity, classification, geometry, topology, and mereology through examples based on the imaginary island state of the Merea. By the end of the chapter, you will be familiar with common issues and how to identify and represent some of the more common forms of GI. Those with little knowledge of GI will be more aware of its nature, and those new to Linked Data will have a better understanding about how it should be approached. The chapter paves the way for Chapters 7 and 8 concerning the publication and linking of Resource Description Framework (RDF) datasets and identifies certain issues that cannot be fully resolved in RDF and RDFS (RDF Schema) but can be better addressed by techniques using the OWL (Web Ontology Language) ontology language and described in Chapter 10.

6.2 IDENTITY: DESIGNING AND APPLYING UNIVERSAL RESOURCE IDENTIFIERS

We begin our story from the perspective of Merea's national mapping agency, Merea Maps. When considering identity, Merea Maps needs to think about what it is interested in identifying with URIs (Uniform Resource Identifiers) and then think about how to construct these URIs. Its two main products are a place-name gazetteer and a detailed digital topographic map. We begin with the gazetteer, which identifies all the named entities represented on the topographic map: settlements; sites and other places (e.g., factories, hospitals, farms, etc.); buildings; roads; rivers; hills; valleys; and so on. All these will need to be assigned a unique URI, as will any necessary properties between them. The gazetteer is required to support a few topological and mereological properties, such as "within" and "part of," like "Medina is within North Merea" and "Ash Fleet Farm House is part of Ash Fleet Farm."

Let us first consider the identification of "things," Merea Map's real-world features, such as buildings, hills, and so on. The URI has essentially two components: something to identify the domain and something to identify the thing within the domain. The domain is the easy bit; if we suppose that Merea Maps has the domain www.mereamaps.gov.me, then it could simply choose to use this. More typically, though, organizations tend to make explicit the fact that the domain is about identifying things and so chooses the domain name: id.mereamaps.gov.me. So, we now have the first part of our URI. Designing a template for the remaining part is more contentious, and there are two main schools of thought. The debate is around how human readable the URI should be. We will use the town of Medina and Ash Fleet Farm as examples. If we wish to make the URI very human readable, we might devise schemes such as <domain>/<type of thing>/<name of thing>, such as

```
http://id.mereamaps.gov.me/town/medina
```

and

```
http://id.mereamaps.gov.me/farm/ashfleetfarm
```

The advantage of this method is that it is possible to immediately tell what the URI is referring to from the URI. But, the method has two major disadvantages; one concerns change and the other uniqueness. The first problem is exposed by the simple question: What happens if the thing changes its type or changes its name (or the type or name was incorrectly specified in the first place)? So, what would happen if Medina became a city? And, what if Ash Fleet Farm becomes Lee Farm following a change of ownership? Ash Fleet Farm was once called Manor Farm, so how would you reference historical data concerning Manor Farm? There are two ways to handle this. We may decide that a change of classification or a change of name means that we have a new thing to describe, so we would allocate a new URI for this new thing. However, geography is rarely that simple, and if no other changes have taken place, more likely than not people will still consider it to be the same original thing. At the very least, it is impossible to be consistent in a way that matches people's expectations. So, this might work in some cases but is unlikely to work in all and would be an inappropriate solution if the original name or classification was simply wrong and the change a correction. The only other option is to live with the old URI and accept that it is now misleading, but this rather undermines the purpose of trying to make the URI human readable.

The second problem, that of uniqueness, is no easier to resolve. The problem is that we cannot guarantee that names are unique, and indeed our experience tells us that they are not. What if there is another Ash Fleet Farm? Let us suppose that one farm is in North Merea and the other in South Merea. Well, we could differentiate them by including their region:

```
mm:/farm/southmerea/ashfleetfarm and
mm:/farm/northmerea/ashfleetfarm
```

Here "mm:" is used to abbreviate the namespace http://id.mereamaps.gov.me, a shorthand that is used from this point. But again, this does not provide a complete or elegant solution. The problem is that even this cannot guarantee uniqueness: What if there are two Ash Fleet Farms in South Merea? In many areas, it is common to find the same names being used to identify different things in very near proximity; for example, in the English county of Wiltshire, there are two villages called Fyfield within 6 miles of each other. It is therefore difficult to devise a system that can guarantee that the URI will not be either ambiguous or so unwieldy to be effectively unusable. To be more precise, in most cases it is impossible.

An alternative approach that resolves the issues of both change and uniqueness is simply to make the URI opaque, that is, not even try to make it human readable. This is our view and was also the view of Tim Berners-Lee when considering URIs in the more general context of the Web as a whole: "The only thing you can use an identifier for is to refer to an object. When you are not dereferencing, you should not look at the contents of the URI string to gain other information" (Berners-Lee, 1996). In this view, the URI is usually based on a combination of domain name and some arbitrary unique ID, normally just a number that is assigned to the thing being identified and never reissued to anything else. So, in such a scheme Medina could have a URI of the form mm:/670020 and Ash Fleet Farm would have mm:/871113. Such URIs are quite jarring to the eye, but they are nonetheless free from the problems identified. Small compromises are possible; for example, we could differentiate between physical things and administrative areas so that the farm could be identified as a topographic object: mm:/topo/871113. The ugliness of this solution is likely to become less of an issue in the fullness of time as better Linked Data tools are used, the URI becomes less visible, and it is less relevant whether it is visible or not.

Last, we have spoken up to this point only of the situation from the perspective of Merea Maps. If we consider the situation from the point of view of a different organization, then that organization may choose to use the URIs allocated by Merea Maps or allocate its own URIs to the same things. This of course means that while URIs must be unique, they need not be exclusive. Simplistically, the ideal is for each thing to have only one URI assigned, but this is rarely possible, so we have to accept that things will have multiple URIs assigned by different organizations. The reasons why an exclusive URI is not always possible are twofold. It is necessary for the second organization to know that a URI will always have been assigned by the first organization and that this URI is publicly accessible; this is rarely the case: There is always the possibility that the receiving organization will be in a position to acquire information about the existence of the resource before the first organization. There are occasions when this will definitely not be the case, a postal service issuing postcodes, for example, but this kind of guarantee is rare. The second issue is one of trust: How much does organization A trust organization B to always have the URI available? This problem is likely to be reduced with time as organizations become better at managing URIs, but it is always likely to exist to some extent or other.

The trick therefore becomes how to manage the problem of multiple URIs rather than attempt to resolve it. Mechanically, this is about correlating these different URIs using `owl:sameAs`, but this is really an abbreviation for the general problem of data conflation, something that will be dealt with more fully elsewhere in this book.

In summary, the main issues when considering assignment of URIs are whether it is possible to reuse URIs already assigned by others and, if a new URI is needed, how opaque it should be, that is, how much of the semantic meaning of the item should be encoded in the URI itself. The former problem is one that revolves around the mechanics of the order of assignment and trust. The second problem is one of how reliable and resilient to change a URI can be if it attempts to embed semantics of the resource within it.

6.3 IDENTITY: NAMES

Names, whether they are place names or names of any other thing, can be complicated. Place names are also known by the formal term *toponym*, but we use place name in this book. Places can have any number of names: an official name (sometimes more than one), short names, vernacular names (ones given through common usage that may or may not be official), historic names, and so on. Names may not be unique: think of how many High Streets and Springfields there are. Indeed, names are rarely unique. Names also can have their own life cycles, quite independent of the thing that they name. In our example island of Merea, Ash Fleet Farm was once known as Manor Farm, a name not used for over 100 years. So, we might want to hold information about the name itself, not just about the thing it names.

But, let us start with a simple case and then progressively look at the possible complications; like many things, the level of complication that you choose to represent will be dependent on the complexity of the situation you encounter and the need you have to represent that complexity. A general piece of advice is to keep it as simple as possible. The simplest possible case is when a place only has one name and will only ever have one name (or you will only ever care about one name). In this case, the simplest thing to do is to represent the name as a string, for example, "Ash Fleet Farm." Now, even in this simplest example there are choices to make. These choices concern the use of the relationship (property) that is used to associate the name with the place it identifies. The most obvious thing to do is to use `rdfs:label`, as discussed in Chapter 5. After all, is this not what this property is all about, to provide a human-readable label for the thing? The answer is yes, but; the "but" concerns technical limitations that apply to this property and most specifically the fact that you cannot define a domain or range for this property. Also, the established use of `rdfs:label` is to be a simple human-readable label for anything (things, properties, and datatypes), whereas we would like to assign the semantic meaning of "name" to our property. It is therefore better to define a separate property that you have complete control over, or to use an already-established property, the most obvious being `foaf:name`. The use of `foaf:name` is a popular solution for naming places, but it was really designed to identify people, not places. However, there is nothing at present to limit the use of `foaf:name` to people, so in many respects it is fine to use `foaf:name`; after all, why invent when you can reuse? The arguments against this course of action are not overwhelming but nonetheless should be taken into consideration. The first question is simply a matter of how comfortable you feel about reusing something not really intended for the purpose you have. The second is potentially more damaging (although probably unlikely in this case): What do you do if the World Wide Web Consortium (W3C) decides to give `foaf:name` the domain of `foaf:Person`? Now, wherever you have used `foaf:name` the implication is that the named places are also people. The safest solution is the establishment of a simple property to name places, such as "has place name." Until such a property is established as a standard, it is probably better either to use `foaf:name` or to define your own. Perhaps a reasonable compromise is to define your own place-name property with a domain of places and make it a subproperty of `foaf:name` as follows:

The relationship "has place name"   mm:hasPlaceName :subpropertyOf
  is a special type of the   foaf:name.
  relationship "name [foaf]".   mm:hasPlaceName :domain mm:Place.
The Relationship "has place name"   mm:/871113 mm:hasPlaceName
  must have the subject Place.   "Ash Fleet Farm".
Ash Fleet Farm URI mm:/871113.
Ash Fleet Farm has place name
  "Ash Fleet Farm".

Then, if the worst comes to the worst and foaf:name is constrained by the domain of foaf:Person, you will only have to break the link between the two properties. We can start to handle further levels of detail by generating new subproperties for different types of place name that may be applied to the farm, for example, old names, shortened forms, and so on:

Ash Fleet Farm has old name "Manor Farm".
Ash Fleet Farm has preferred name "Ash Fleet Farm"[1].
Ash Fleet Farm has short name "Ash Farm".
Ash Fleet Farm has colloquial name "Ashy Farm".

And so on.

Each needs to be made a subproperty of mm:placename, which means each will also inherit any domain or range restrictions, so these do not need to be repeated. So, to define and use the property hasOldName property, all we need to do is

The relationship "has old name" is a   mm:hasOldName :subpropertyOf
  special kind of the relationship   mm:hasPlacename.
  "has place name".   mm:/871113 mm:hasOldName
Ash Fleet farm has old name   "Manor Farm".
  "Manor Farm".

You might ask why it is useful to make these all subproperties of mm:hasPlaceName. The reason is that by doing so you can ask the question: What place names are associated with Ash Fleet Farm? and the query will return all of these names. So, we now have an easy way to return all the names: place name (current name), old names, short names, and so on.

But, things can be more complicated than this. What if we want to say that the name Ash Fleet Farm only became the name of the farm in 1891 when it was renamed from Manor Farm? Or similarly, that Manor Farm became the old name for Ash Fleet Farm in 1891? We can do this without changing anything that we have already put in place by adding further properties to our ontology. For example, "has a preferred name valid from" and "has an old name valid from":

Ash Fleet Farm has a preferred name valid from 1891.
Ash Fleet Farm has an old name valid from 1891.

But, this only works in very limited circumstances. What if there are multiple old names? To which old name does the previous statement refer? There is no way

of knowing. The problem is that these are statements that should really apply to the
name, not the place. The way to get around this is to treat the name as a resource,
not just a string value. This means we give it its own URI and can then say as much
about it as we like. For example, we can state the following:

```
Ash Fleet Farm has place name        mm:/181113 mm:hasPlacename mm:/n200.
 Ash Fleet Farm (name).²              mm:/n200 mm:text "Ash Down Farm".
Ash Fleet Farm (name) text           mm:/n200 mm:firstUsed 1891.
 "Ash Fleet Farm".                   mm:/n200 mm:supersedes mm:/n305.
Ash Fleet Farm (name) first
 used 1891.
Ash Fleet Farm (name)
 supersedes Manor Farm (name).
```

Here we are saying that Ash Fleet Farm has a name, and that the name itself is
important, and we can say things about it in its own right. This system is more complicated than just treating the name as a value associated with the place, but we trade
off complexity for additional expressivity. One loss is that it is no longer possible
to associate the naming mechanism with the widely used foaf:name as it is too
simple an implementation. An advantage for some people is that the name is more
important, or at least as important, as the thing it is naming. People who need this
kind of use case typically might work for heritage organizations for which the name
is of significant historic interest. So, treating the name as a thing in its own right and
assigning it a URI not only enables Merea Maps to be more expressive, but also helps
its sister government agency Merea Heritage.

Names are therefore not always as simple as we would like them to be, and we
have to be careful when representing them as Linked Data. On one hand, we do not
want to make things too complicated, but on the other we want to make sure we can
represent them properly and at a level that is appropriate for our end users.

6.4 GEOMETRY

We have seen previously that geometry is a very important attribute of geographic
objects, and that in vector terms they are typically represented as points, lines, and
polygons. The simplest case is where a feature is represented by a single point. This
is certainly as simple as it gets, but even here choices exist concerning the projection system used. WGS86 is the most commonly used encoding of geometry on the
Web, which expresses the latitude and longitude of a feature, using two properties
Lat and Long. So, we could say that the Lat/Long of a representative point for Ash
Fleet Farm could be

```
Ash Fleet Farm Lat 45.19964.        mm:/181113 Lat 45.19964.
Ash Fleet Farm Long 5.65749.        mm:/181113 Long 5.65749.
```

Other coordinate systems can be applied by using alternate or additional properties. For example, Merea has a local grid system and tends to publish data with both
the local X,Y grid and Lat/Long.

```
Ash Fleet Farm X 4875.            mm:/181113 X 4875.
Ash Fleet Farm Y 1287.            mm:/181113 Y 1287.
```

This simple system works well enough if a single point is required. If multiple points are needed, then these simple properties are insufficient because it would not be possible to determine which Lat corresponds to which Long. This also raises the question of just how useful it is to represent geometry as pure RDF. Consider how you might represent a line or polygon as a series of linked triples. It can certainly be done, but to what advantage? It would only be necessary on the rare occasion when it is required to associate further unique data to an individual coordinate. The alternative is to bundle up all the coordinates in some structured form so that it can be unbundled when it is needed. Fortunately, the Open Geospatial Consortium (OGC) has already done this for us as an extension to the standard datatypes that are supported by RDF and defined as part of their GeoSPARQL specification (Perry and Herring, 2011). We will not go into all the nitty-gritty here, but focus on the essentials of how the geometry is associated with the feature being described and then how the geometry is represented. Using GeoSPARQL for querying spatial data is discussed more in Section 8.3.

The OGC Simple Feature model represents a spatial object as an OGC feature, which is essentially an abstract representation of a real-world object. The OGC feature is assigned a URI, and the property geo:defaultGeometry[3] is used to associate the geometry to the OGC feature. The namespace geo: expands to http://www.opengis.net/ont/OGC-GeoSPARQL/1.0/.

The geometry itself is represented using WKT (Well-Known Text) as defined by the Simple Feature model International Organization for Standardization (ISO) 19125-1 (ISO 19125-1 2004). The WKT format (or serialization) represents points, lines, and polygons as follows:

```
Point     Point (x, y)
Line      Linestring (x1, y1, x2, y2, ... xn, yn)
Polygon   Polygon ((x1, y1, x2, y2, ... x1, y1))
   and    Polygon ((x1, y1, x2, y2, ... x1, y1), (a1, b1, ... an, bn, a1, b1))
```

The polygon has two formats, the first for simple polygons and the second for polygons with holes, each hole being represented by a separate closed linestring. Each linestring is closed by repeating the first coordinate pair as the last. The following code snippet demonstrates how a WKT polygon can be associated with some feature (featureX) in RDF:

```
1. <example:feature rdf:about = "example:featureX">
2.   <geo:defaultGeometry rdf:resource = "example:geometry"/>
3. </example:feature>
4.
5. <ogc:Polygon rdf:about = "example:geometry">
6.   <ogc:asWKT rdf:datatype = "http://www.opengis.net/rdf#WKTLiteral">
7.   http://www.opengis.net/def/crs/OGC/1.3/CRS84
8.     Polygon((5.6 20.2, 5.6 20.7, 5.3 20.5, 5.3 21.9, 5.6 20.2))
```

```
9.    </ogc:asWKT>
10.</ogc:Polygon>
```

Lines 1 to 3 state that feature is an instance of `example:feature` with a default geometry identified of `example:geometry`. `example:geometry` in turn is an instance of the `ogc:Polygon` class that has the literal value defined by the WKTLiteral (line 6 specifies that the coordinate reference system being used is WGS84). Points and lines can be expressed in a similar fashion.

6.5   CLASSIFICATION

Another important aspect of a geographic feature is what it is, that is, how it is classified. We give a fuller description of how we build classification systems in the chapters on ontologies but start with a brief introduction here.

As we said in a previous chapter, in RDF we can give a thing a classification using the statement `rdf:type`:

```
Ash Fleet Farm is a Farm.
mm:181113 rdf:type mm:Farm.
```

which simply says that Ash Fleet Farm is an instance of a particular class called a Farm. It does not, however, explain what a Farm *is*. We deal with how we provide a description of a class when we discuss the development of ontologies in further chapters. For now, it is sufficient to understand that once we have defined something as a Farm we can ask questions about Farms and get results that include Ash Fleet Farm.

There is nothing to stop us from applying multiple classifications to an object; we could, for example, also define Ash Fleet Farm as a heritage site and a conservation area. There is nothing wrong in doing so, but we need to think carefully about why we might do this. Often when we assign different classifications to the same thing, we are mixing different contexts. In the example case, we are mixing its function (it is a farm), its historic interest (it is a heritage site), and ecology (it is a conservation site). We need to carefully consider whether the whole of the farm is a heritage site or conservation area. If not, we may be really dealing with different areas and hence different things, even though they may contain some reference to Ash Fleet Farm. For example, parts of Ash Fleet Farm, including the main farmhouse, are part of a conservation area known as Ash Fleet Farm Conservation Area. However, since only parts of the farm, and not the entire farm, are included in the conservation area, defining the Ash Fleet Farm as a conservation area would be incorrect. Indeed, what this tells us is that a new feature or resource exists: Ash Fleet Farm Conservation Area, which should have its own URI and be treated independently from the farm.

6.6   TOPOLOGY AND MEREOLOGY

A geometrical description of a geographic object is obviously important, but in Linked Data terms, topology (explicit spatial properties) and mereology (part-whole properties) are probably more important.

6.6.1 Topology

Topology can be used to express the spatial properties between topographic features, for example, whether two buildings are next to each other, and to express the connectivity within a network such as a road system. In the former case, relations can be based on a formal system such as Region Connection Calculus (Randell, Cui, and Cohn, 1992) or the Egenhofer nine-way intersection model (Egenhofer, 1989), or alternatively on a more informal system with less-well-defined semantics.

6.6.1.1 Region Connection Calculus 8 and OGC Properties

Region Connection Calculus 8, or RCC8 as it is frequently abbreviated, is a set of eight topological properties that can exist between two regions or areas. Other systems with fewer or more properties can also be specified (such as RCC5), but RCC8 properties are the most frequently used by Geographic Information Systems (GIS). The specific properties are shown in Figure 6.1.

OGC implemented these properties as shown in Table 6.1,[4] where the geo namespace is http://www.opengis.net/ont/OGC-GeoSPARQL/1.0/.

The names given to the properties are somewhat obscure to say the least and reflect the fact that they are the product of mathematical minds. We can simply rename these for convenience, but OGC also provided an alternative set of properties that can be mapped to the RCC8 relations (as shown in Table 6.2) and that serve most purposes.

A further problem with the RCC8 names is that some imply they are more than topological properties; they are also membership properties: Does a tangential proper part imply that **a** is a *part* of **b** as well as being surrounded by it? From the mathematics, it is clear that the intention is only to express topological properties. The OGC terms more clearly show that it is topology not mereology that is in force.

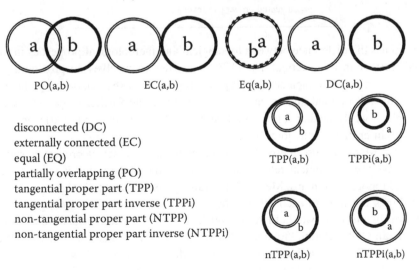

disconnected (DC)
externally connected (EC)
equal (EQ)
partially overlapping (PO)
tangential proper part (TPP)
tangential proper part inverse (TPPi)
non-tangential proper part (NTPP)
non-tangential proper part inverse (NTPPi)

FIGURE 6.1   RCC8 Relations.

TABLE 6.1
OGC RCC8 Properties

Relation Name	OGC Relation URI
equals	`geo:rcc8-eq`
disconnected	`geo:rcc8-dc`
externally connected	`geo:rcc8-ec`
partially overlapping	`geo:rcc8-po`
tangential proper part inverse	`geo:rcc8-tppi`
tangential proper part	`geo:rcc8-tpp`
non-tangential proper part	`geo:rcc8-ntpp`
non-tangential proper part inverse	`geo:rcc8-ntppi`

TABLE 6.2
OGC Properties Mapped to RCC8 Relations

OGC Properties	Relation URI	RCC8
equals	geo:sf-equals	Equals
disjoint	geo:sf-disjoint	Disconnected
intersects	geo:sf-intersects	Not disconnected
touches	geo:sf-touches	Eternally connected
within	geo:sf-within	NTPP + TPP
contains	geo:sf-contains	NTPPi + TPPi
overlaps	geo:sf-overlaps	Partial overlaps

NTPP = non-tangential proper part; NTPPi = non-tangential proper part inverse; TPP = tangential proper part; TPPi = tangential proper part inverse.

We discuss the difference between topological and mereological properties in more detail further in this section. The OGC properties themselves do provide a good basis on which to implement geographic properties as the relationships between them are well defined and so can be used for inferring further properties. So, for example, if we know that A is within B and that B is disjoint (separated) from C, we can also infer that A is disjoint from C.

Merea Maps can use these OGC relations to define a set of properties suitable to express the topological relationships between the objects that it records on its detailed topographic maps. Merea Maps wants to use a "next to" relation to say that two things are next to each other if they physically touch—two cottages side by side within a terrace[5] would therefore touch. This is very straightforward as there is already an OGC relation "touches" that seems to exactly meet these needs.[6] It can use "touches" as is but decide instead to use the term "next to" as they feel this is a more obvious term from the perspective of their users.[7] So, all they have to do is specify that "next to" is a subproperty of "touches" and perhaps define a domain

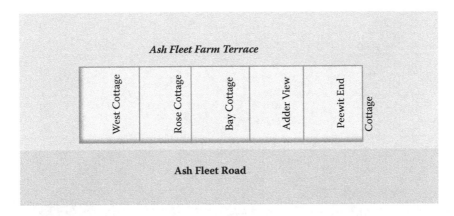

FIGURE 6.2   The Ash Fleet Farm Estate cottages.

and range to say that it refers to topographical objects only. Merea Maps uses the term "near to" rather than "touches" not just so we can specify that it applies to topographical objects but also because the term "touches" is not used in geography.

```
"next to" is special type of "touches" [geo].
"next to" must have the subject "Geographic Object".
"next to" must have the object "Geographic Object".
mm:nextTo rdfs:subpropertyOf geo:sf-touches.
mm:nextTo rdfs:domain mm:GeographicObject.
mm:nextTo rdfs:range mm:GeographicObject.
```

So, if we consider the row of terrace cottages (Figure 6.2) that are part of the Ash Fleet Farm estate, we can say the following:

```
Rose Cottage is next to Bay Cottage.      mm:875483 mm:nextTo mm:875484.
Bay Cottage is next to Adder View.        mm:875484 mm:nextTo mm:875485.
```

and so on. MereaMaps can then define the other relations in a similar way. While we have mentioned that topological properties offer the ability to provide reasoning, we have not yet discussed an implementation of that reasoning; RDFS is not sufficiently expressive to allow complex reasoning, but in Chapter 9 we describe the ontology language OWL that will enable the properties to be more richly characterized.

By using the OGC Relations to explicitly state topological relationships between features, we are able to use these properties to support queries using SPARQL. However, as the properties are grounded in GeoSPARQL, they also enable us to infer the topological properties of the features if there is geometry associated with them.

6.6.1.2   Non-RCC8 or OGC Topology

RCC8 and the OGC subset provide sets of topological properties that can provide certainty in the relationships by being grounded in a well-defined geometry. The problem for geography is that there is more to geography than just geometry. So, while they are a good solution when geography is based on well-defined geometry,

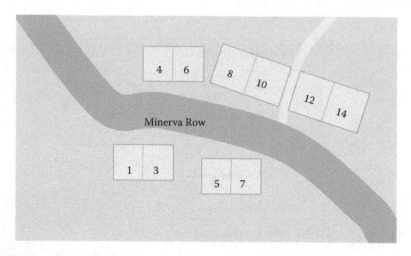

FIGURE 6.3 Isis houses.

they do not provide all the answers. Geography frequently lacks the certainty of hard geometry because precise geometric definitions may be unknown or simply not defined or because people operate at varying levels of detail. People also tend to use a whole host of inexact terms when referring to the spatial properties between two features. Let us look at an example, again using the "next to" property.

We have seen how we can use OGC Geo properties that are based on RCC8 to represent topological relationships, but now consider not the cottages on Ash Fleet Farm estate but some of the houses in the village of Isis (Figure 6.3).

The "next to" property as Merea Maps have currently designed it will work if we are operating at the level of a property (house and gardens) as follows:

```
4 Minerva Row is next to 6 Minerva Row.[8]
8 Minerva Row is next to 10 Minerva Row.
```

But, we cannot say that

```
10 Minerva Row is next to 12 Minerva Row.
```

as there is a path between the two properties, so they are geometrically disjoint. However, not everyone operates at this detailed level, or they may be generally less concerned about precise geometry; Merea Heritage would regard a path as irrelevant, and even Merea Maps when implementing its place name gazetteer will not be interested in the absolute detail of its topographic maps. Similarly, two detached houses each surrounded by adjoined gardens might be considered to be next to each other at one level of resolution: that of the property not the building. It is therefore possible to have a "next to" property as a pure topological property that is not grounded on RCC8. We can define this different "next to" in the same way as we defined the original OGC "touches" version, except that we cannot define it as a subproperty of the geo:sf-touches property. In effect, the main difference is how we use it. But, we have

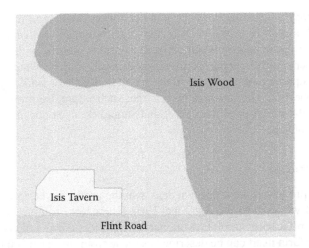

FIGURE 6.4   North of Isis Tavern?

to be careful; the "next to" that is a subproperty of the OGC property is grounded in the GeoSPARQL query language and therefore has a precise implementation; the other definition is entirely dependent on the users knowing the context in which it has been applied. So, when we encounter a "next to" property, we need to check that we understand its semantics: Is it grounded in a well-defined model such as RCC8 or some less-rigorous context? Of course, it is better if the differences are explicitly recognized, so Merea Maps defines a separate property "neighbors" to implement this less-precise property and uses it for their gazetteer. Merea Heritage in turn adopts this property for its own use.

The "next to" property is of course just one of many that can exist in some imprecise form, some of which have RCC8 alternatives and some of which do not. They can include "opposite," "near," and compass directions such as "north of." It is always necessary to be clear about the context in which these properties are used. If we consider "near," then this can be more or less precisely defined for a particular set of circumstances; for instance, we can define "near" as meaning anything within 50 m. The problem with "near" is that it is so context dependent that there can be dozens of different definitions, each valid for a particular circumstance, our "next to" issue writ large. Similarly, there can be problems with compass properties such as "north of," "south of," and so on. Here, the problem is clear to see in Figure 6.4. Is the Isis Wood to the north of the Isis Tavern? Yes, but parts clearly are not and are to the east, but equally saying it is to the east is just as imprecise.

Of course, there is nothing to stop us saying

```
Isis Wood is north of Isis Tavern.
Isis Wood is east of Isis Tavern.
```

The point is that there are very many topological properties that are possible and many different interpretations of these properties depending on the context, such as we have seen with "near to." One could decide that such properties are more trouble than they are worth. However, it is not always possible to rely on geometrical grounded properties either because the exact boundary between two features is unknown, or the boundary is not defined. Consider suburbs within a city. In the United Kingdom, such boundaries are rarely defined, so if we want to express any topological properties we cannot use those based on RCC8 or similar geometrically grounded properties. The message here is that such imprecise properties are very often useful, should be used with care, and should be used consistently, within a well-defined context.

6.6.2 MEREOLOGY

Mereological properties are those that deal with the properties concerned with parts and wholes. If we consider the Ash Fleet Farm Estate, for example, we can say that it is made from (at least) the following parts: Ash Fleet Farm and Ash Fleet Farm Cottages. The farm itself can be described as made from the various farm buildings, fields, woodland, and tracks that make up the farm, and the description is similar for the Ash Fleet Farm Cottages. So, Ash Fleet Farm Estate can be described as made from a hierarchy of parts. In fact, if we so wished, we could describe the whole of Merea as a hierarchical breakdown of the *geographical parts* that comprise it. However, for two very good reasons such extreme hierarchical decompositions are not advisable. First, the hierarchy itself will be very large and unmanageable. Second, although things like Ash Down Farm Estate can be reliably decomposed, larger areas usually present more of a challenge. This is because something like Ash Down Farm Estate is a well-defined entity, and through ownership it is reasonable to think of it as a number of parts that are all owned by the Estate. When we consider larger areas, things become less clear. In Merea, the national government can be said to "own" the island, but in the Merean constitution, the government funds rather than owns the local government. Therefore, while we can describe the topological property between Merea as whole, and its two main local administrative districts:

```
Republic of Merea contains City of Medina.
Republic of Merea contains District of South Merea.
```

we cannot say

```
Republic of Merea has part City of Medina.
Republic of Merea has part District of South Merea.
```

at least not from an administrative point of view. We cannot build up a mereological description of Merea based on the administrative areas. What we can say is that

```
Merea has part Medina.
Merea has part South Merea.
```

Here, we do not refer to the administrative areas but geographical ones. However, even here there will be problems. If we look below the level of South Merea, it is not clear whether Ash Fleet Farm Estate is part of the village of Ash Fleet or not. It is within the parish, but the parish is an administrative division, and we should not mix geographies. Put simply, in the vernacular, the farm is ambiguously defined as either part of the village or not depending on differing viewpoints. Further, the reason for the ambiguity is often because it simply does not matter that much; if it did, people would have resolved the issue. Another problem is that if we do take a geographical approach, then do we include hills, valleys, and other natural topographic features? And if we do, how do we relate the mereological to the man-made features? These questions can only be properly answered if the context is known, but again there is a danger of mixing different geographies. Therefore, we can conclude that the overuse of merelogical properties not only is unwieldy but also exposes all sorts of issues that are difficult to model consistently and offer little advantage to the modeler. So, within geography what we frequently find is that mereological properties are most useful for defining the makeup of well-defined places such as farms, hospitals, and so on. In doing so, it is also important that we should ensure that the description reflects a single discrete geography, and that we do not confuse topological and mereological properties[9] or indeed other properties such as ownership.

So, how should we represent mereological properties? Unlike topological properties, by and large mereological properties are simpler. In fact, the two used most will be "part of" and "comprises" or "has part." We define these much as we defined "next to"; again, we cannot express this using RDFS alone. Most specifically, what is often useful is to be able to infer that if A is part of B and B is part of C, then A is also part of C. How this can be achieved is described in Chapter 10, which describes the application of the OWL ontology language to the domain of geography.

6.6.3 Network Topology

Network topology is the expression of properties that exist between elements of a network of some form or other. In geography, the most common networks are those of roads, rail, and rivers. As Linked Data is structurally a graph (or network), it is able to naturally represent network topology.

In terms of the topological properties themselves, they are mostly concerned with connectivity (Is A directly connected to B?) and flow (If A is connected to B, can I travel from A to B in both directions or not?). Other properties can affect the ease of flow (flow rate, travel times, etc.; the nature of the connection), whether it is a metaled road, a bifurcating stream, and so on and perhaps things like the number of connections in the case of multilane roads or multitrack railways. We initially use the road network between Medina and the village of Isis as an example (Figure 6.5). Here, there are two roads that link the two settlements: the Old Medina Road that goes via the hamlet of Norton Soke; and the newer Medina Road that goes directly to Isis.

The simplest way to express the fact that you can travel between the two settlements is to say something like the following:

```
Medina is connected to Isis.
```

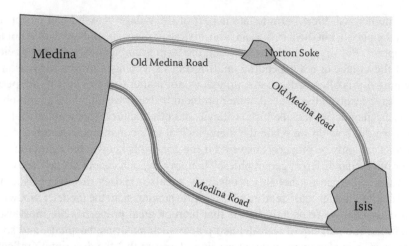

FIGURE 6.5   The Medina Isis Road network.

We can of course also say

```
Medina is connected to Norton Soke.
Norton Soke is connected to Isis.
```

At this point, we are saying nothing about how the settlements are connected; we describe the general connectivity, but not the road network that implements it. We may not need to, in which case this solution is adequate, but should there be a need to do so, then we can say the following:

```
Medina Road is connected to Medina.
Medina Road is connected to Isis.
```

And

```
Old Medina Road is connected to Medina.
Old Medina Road is connected to Norton Soke.
Old Medina Road is connected to Isis.
```

Even this is incomplete for two reasons. The connections as stated only apply in one direction; we know the Medina Road is connected to Medina, but from this statement, we cannot say whether Medina connects to the Medina Road. If we are just using RDF and RDFS, the only way we can handle this is to produce a mirror statement that explicitly states

```
Medina is connected to Medina Road.
```

In Chapter 10, we discuss a better solution using OWL to assign additional characteristics to the "is connected to" property. For now, let us assume that all further examples in this chapter are mirrored in both directions. Of course, if a connection is genuinely only in one direction, such as would be the case with a one-way road, then mirroring is inappropriate.

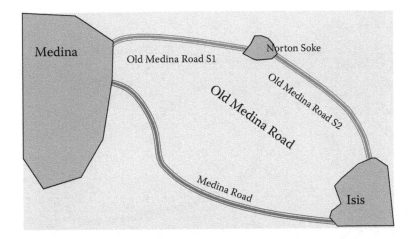

FIGURE 6.6   Dividing the Old Medina Road into sections.

The second problem with our statements is that they do not provide enough information on their own to reconstruct the network. Does the order go Medina, Norton Soke, Isis, or Norton Soke, Isis, Medina, or what?

To get around this, it is more normal to break the road up into sections as shown in Figure 6.6 and then use these sections to connect the settlements:

```
Old Medina Road S1 is part of Old Medina Road.
Old Medina Road S2 is part of Old Medina Road.
Old Medina Road S1 is connected to Old Medina Road s2.¹⁰
Old Medina Road S1 is connected to Medina.
Old Medina Road S1 is connected to Norton Soke.
Old Medina Road S2 is connected to Norton Soke.
Old Medina Road S2 is connected to Isis.
```

It is now possible to reconstruct the connectivity.

In fact, it is normal to decompose a network connection for two main reasons: to enable the network connectivity to be correctly expressed as in our example and to enable properties to be associated with specific sections of the connection. In the case of the Old Medina Road, one section of the road between Norton Soke and Isis is single track, and we want to record this explicitly. To do so, we further break down the road into four sections rather than two, as depicted in Figure 6.7.

The route between Norton Soke and Isis is then described as follows:

```
Old Medina Road S2 is connected to Norton Soke.
Old Medina Road S2 is connected to Old Medina Road S3.
Old Medina Road S3 is connected to Old Medina Road S4.
Old Medina Road S4 is connected to Isis.
Old Medina Road S2 has exactly 2 Lanes.
Old Medina Road S3 has exactly 1 Lane.
Old Medina Road S4 has exactly 2 Lanes.
```

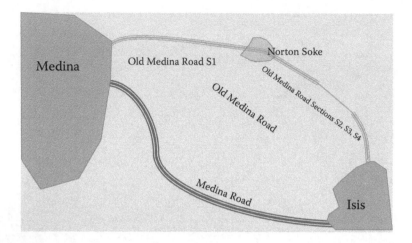

FIGURE 6.7   Decomposing the network further.

However, it is not always sensible to do this. In this case, it is likely that the situation as described will remain as it is for some time to come. If the single-track section is upgraded at some time in the future, we could either merge the road sections back together or just leave things as they are but attribute S3 with having two lanes—the actual solution will depend on how you want to manage these life cycles. But, what if the attribute we are concerned with is the position of road signs? In this case, there is a lot of potential for the roads to become oversegmented and for relatively frequent changes in the number and position of road signs to generate significant extra work and change. In this case, it is better to model the road signs as objects in their own right (as they are) and to link them to the road segment:

```
Sign M12 has Lat 45.19847.
Sign M12 has Long 5.65332.
Sign M12 has text "Single Track Road Ahead".
Sign M12 is on Old Medina Road S4.
Sign M12 faces towards Norton Soke.
```

To record the extent of a road restriction, such as the applicable speed limit, we can record start and end points, plus the limit:

```
Speed Restriction 2765 applies to Old Medina Road S4.
Speed Restriction 2765 begins at latitude 45.19659.
Speed Restriction 2765 begins at longitude 5.65347.
Speed Restriction 2765 ends at latitude 45.19661.
Speed Restriction 2765 ends at longitude 5.65344.
Speed Restriction 2765 restricts max speed to 50.
```

6.6.3.1   Links and Node Network Model

Readers familiar with networks may be surprised that there has been no mention of the classical link-and-node model for networks. In this model, all links connect to nodes, and no link may connect directly to another link but must do so via a node. It is fairly straightforward to represent this model as we can generate link and node classes and map our Features to them. However, it does raise an interesting issue: In Linked Data terms, the things we represent are real things that we wish to describe using information that is obtained by dereferencing the associated URIs. However, the classical network model is a representation of an *abstraction* of the real-world things. To explain this difference further, consider how we would represent our example network. Medina is a City (in turn a kind of settlement, that is in turn a Feature), and Medina Road is a road (in turn a Feature), so we would expect these assertions to have been explicitly stated:

```
Medina is a City.
Medina Road is Road.
```

However, in our abstract network model, Medina is also a Node, and Medina Road a Link:

```
Medina is a Node.
Medina Road is a Link.
```

So, now Medina is both a City and a Node, but can it actually be both in the same model? One is modeling a real-world thing, the other an abstraction. We also saw the Old Medina Road is broken up into a number of road stretches, and these are directly connected. In the network model, we cannot do this; they have to be connected via nodes, so we have to invent nodes to join the road stretches. Now, it can be argued that Merea Maps also had to invent road stretches, but in this case, it is easy to demonstrate the physical existence of the stretches. It is very common practice for roads to be segmented in this way and the stretches named for road maintenance and management purposes; of course, in many cases these may be considered and named as roads in their own right. The same cannot be said for the invented nodes; they physically represent nothing in the real world; they are a requirement of an abstract model and nothing more. We could change the relationship from "is a" to "is represented by," but the real point is that the abstract network model is not really adding anything other than additional complexity. This is not to say that this model is not without its benefits: It provides a uniform way of representing a network that can significantly improve computational efficiency over the network. However, this example highlights a difference in purpose between the two ways of representing a network: The first focuses on simplicity and accuracy of representation, the latter on consistency and computational efficiency. Each therefore makes compromises, and each is appropriate in its place. It is also worth adding that it is perfectly simple to represent the standard link and node model as Linked Data, and there will be many occasions when this is what you need to do. For example, it is better to express the network in link and node form if you know in advance that the network will be used for routing.

The general issue is of course not a problem specific to networks; it will occur wherever an abstract model introduces modeling elements that have no real-world counterpart. The OGC Feature and INSPIRE Spatial Object are other examples of this.

6.7 SUMMARY

This chapter has shown how GI may be represented as Linked Data. It has discussed identity in terms of both digital identity (URIs) and place names, as well as covering basic classification and shown how topological and mereological relationships can be handled. It has also highlighted how there are always choices to be made, and that the modeler is always faced with balancing conflicting interests—most often between model richness and complexity. Understanding end-user requirements and minimizing complexity are the most important factors to take into account when representing GI as Linked Data.

NOTES

1. This is your preference, no one else's.
2. For brevity, from now on we will omit the statement that maps a succinct name to a URI, in this case: Ash Down farm (name) has URI mm:/181113.
3. The model allows multiple geometries to be associated with the OGC Feature; here, we only deal with the default. The representation of other geometries is the same.
4. For completeness, OGC also implemented the Egenhofer nine-way intersection model.
5. In the United Kingdom and Merea, one use of the term *terrace* is to describe a row of houses or cottages that are physically constructed as a single building, and hence each house or cottage shares at least one wall with another.
6. There is one caveat that needs to be made clear here: The range and domain of the OGC properties are geo:SpatialObject. If you do not want your features to be inferred to belong to this class, then you will have to implement your own properties.
7. If Merea Maps could be sure that semantic Web applications using their data would only present the rdfs:label to the end user, an alternative here would be to encode the term intended for human consumption, "next to," in the rdfs:label instead.
8. From here, we only use the Rabbit notation to avoid needless repetition unless the use of Turtle is helpful to demonstrate some aspect of the syntax not previously introduced.
9. There are occasions when the topological and mereological properties are strongly connected; this is explicitly recognized as *mereotopology*, for which the properties imply both part-whole and topological properties.
10. This statement is only required if you are interested in both the connectivity between roads and the connectivity between places.

7 Publishing Linked Data

7.1 INTRODUCTION

In Chapter 5, we introduced the Resource Description Framework (RDF), the World Wide Web Consortium (W3C) standard markup language for encoding Linked Data, and in Chapter 6, we demonstrated how Geographic Information (GI) could be modeled in RDF. In this chapter, we explain how to take the GI you have, which may be in a relational database (RDB), GIS (Geographic Information Systems), or stored as XML (eXtensible Markup Language) files based on Geography Markup Language (GML), and convert it to RDF for publication as Linked Data. The purpose of publishing your GI as Linked Data is so that it can be more easily reused. This chapter first states the main principles of Linked Data, as originally set out by Tim Berners-Lee (2006), then discusses how to design an RDF data model based on the structure of your current GI. We also look at the tools available for storing, publishing, and querying your RDF data, as well as introduce methods for including metadata about where the data has come from (its provenance) and its licensing terms.

Consideration must also be given to linking your dataset to others on the Web of Data, so that it can be easily discovered. Hence, in Chapter 8 we explain the mechanisms for selecting the most appropriate links between your data and others', which in effect is addressing the problem of data integration.

7.2 LINKED DATA PRINCIPLES

The four Linked Data Principles as set out by Berners-Lee in his Linked Data Design Note (Berners-Lee, 2006) are as follows:

1. Use Uniform Resource Identifiers (URIs) as names for things.
2. Use Hypertext Transfer Protocol (HTTP) URIs so that people can look up those names.
3. When someone looks up a URI, provide useful information, using the standards (RDF, SPARQL).
4. Include links to other URIs so that they can discover more things.

The first point is self-explanatory; we have already covered the use of URIs to name things in the real world, and that we need to assign URIs for the things themselves separate from the URIs assigned to the Web resources describing those things. The second point is again a technique that we are already using: Merea Map's URIs are of the form http://id.mereamaps.gov.me/something. But, it is worth explaining why Tim Berners-Lee thought it was so important to have URIs based on HTTP. HTTP URIs mean that the URI can be looked up on the Web, and we can retrieve the description of the object; this is known as a *dereferenceable* URI. The third principle states that when we retrieve the description of the object, we should be able to understand it, so we use a standard language like RDF to encode that description. The fourth principle recommends adding links between different datasets to be able to move along these links from one dataset to another. This then positions each dataset as a connected subpart of the whole Linked Data Web and enables applications such as semantic search engines to crawl along these links and discover data across different datasets. Furthermore, as a link is encoded as the predicate of an RDF triple, it is more than just a Web hyperlink, as it is typed, and hence has meaning in itself.

7.3 MAKING URIS DEREFERENCEABLE OR SLASH VERSUS HASH

There are two ways to dereference a URI, that is, to retrieve the description of the actual object we are talking about, using HTTP. One is to use a 303 URI, which will be of the form http://id.mereamaps.gov.me/topo/00012, a "Slash URI", and the other is to use a Hash[1] URI of the form http://id.mereamaps.gov.me/topo#00012. You will see both mechanisms used, but they have slightly different effects, and you should consider carefully, based on your particular use case, which is more appropriate.

7.3.1 SLASH OR 303 URIs

The Slash or 303 URIs are so called because when the client requests a URI of the form http://id.mereamaps.gov.me/topo/00012 using an HTTP GET command, the HTTP server responds with a code "303 See Other" plus the URI of a Web document that describes the object, for example, http://data.merea.gov.me/topo/00012. As suggested by Heath and Bizer (2011), it can be helpful to follow the convention of using "id"[2] in the URI when referring to the resource (real-world object) itself, "page" when specifying a HTML Web page, and "data" when referring to the RDF/XML document describing the resource. This means that the client is redirected to this second URI, and now the client also needs to dereference it to get to the Web document describing the object. So, the client then sends a second HTTP GET request on the second URI, and the server replies with an HTTP response code 200 OK plus the RDF document describing the original object. This somewhat convoluted process of redirection means that the server is forced to do the hard work of finding the individual resource (in the second query) rather than leaving it to the client to do so. The advantage of this system is that the whole URI can be retrieved from the server.

7.3.2 HASH URIs

The drawback of the 303 or Slash URI method is that the client has to make two requests, and the server has to respond twice. An alternative is to create URIs using hashes (#) in the form http://id.mereamaps.gov.me/topo#00012. In this case, the HTTP protocol will strip away the bit after the #, which is known as the fragment identifier, and then only the part before the # is requested from the server. This means that the URI as a whole cannot be directly retrieved from the server, so there might not be a

Web document describing the object available at all. For our example, an HTTP GET request is made for http://id.mereamaps.gov.me/topo. The server then returns the whole RDF/XML file for all the Topographic Object data, and the client then has to run its own code to search for triples involving http://id.mereamaps.gov.me/topo#00012.

7.3.3 SLASH VERSUS HASH

The advantage of the Hash mechanism is that only one HTTP GET request needs to be made, so data access is faster. However, the downside is that all RDF triples that share the same URI up to the # will be returned, which could be a big overhead if the dataset is large. We would recommend using 303 redirects (Slash URIs) when the dataset to be queried is large, particularly when the application does not demand many repeated queries. Hash URIs are more suitable for early testing or prototyping purposes, as the server does not need to be configured to perform the 303 redirects, or when the RDF dataset is small and unlikely to change much over time. Hash URIs are also used when RDF is embedded in HTML Web pages using RDFa (Resource Description Framework–in–attributes), helping to keep the HTML document's URI separate from the URIs of the RDF resources.

7.4 LINKED DATA DESIGN

This section assumes that you have some GI stored in a GIS or RDB or encoded in an XML-based GML file or similar output format, and your aim is to publish it as Linked Data. If you already have RDF data in a triple store, and you are satisfied with your RDF data model, skip forward to Section 7.5.5., where discussion on publishing RDF data begins.

There are five main steps in the process of designing Linked Data: First, decide what your Linked Data is about; second, look at your current GI data; third, specify your RDFS (RDF Schema) ontology[3]; fourth, mint create your URIs; and finally, generate your Linked Data. We look at each of these steps now in turn.

7.4.1 STEP 1: DECIDE WHAT YOUR LINKED DATA IS ABOUT

We go into more detail about how to design an ontology in Chapter 10, but the process for developing an RDFS ontology follows along the same lines, albeit producing a simpler structure. First, think about the purpose of your data. What will you use it for? And, a more difficult question: What might other people want to use it for in the future? While you can never plan for all the possible future uses, selecting meaningful names that are well known in your subject area and are not limited to your own organization's jargon will help with this. Also, avoid overspecifying your data in the RDFS ontology. If you include domains and ranges for every class, this will limit their future reuse. Make sure that everything included in the RDFS ontology is absolutely necessary to describe your classes. Also, make a statement of the scope: What should be included, and as important, what *is not* necessary to be included in the RDFS ontology? Scope creep is an inevitable pitfall of authoring an ontology, and having an explicit stated scope will help you avoid trying to describe the whole world. The third initializing step in schema creation is to select what are known as competency questions. These are in effect your test set: the questions that the RDFS ontology and accompanying data must be able to answer correctly. Further in the process, these will be translated into the RDF query language SPARQL, and the Linked Data set queried, to make sure it fulfills your requirements.

If Merea Maps wants to create some Linked Data about its administrative regions, it could state the purpose of its Linked Data to be: "To describe the administrative geography of Merea and the topological relationships between them." The scope would be: "All levels of administrative area that occur in Merea, their sizes, point locations, the topological relationships between areas of the same type. Authorities that administer the regions are not included, nor are the spatial footprints of the regions," and they could have a set of competency questions, which would include, among others:

- Name all the Counties in Merea.
- Find all the Cities in Merea.
- Which Parishes are in Medina County?
- What is the latitude/longitude of Medina City?
- Find the Counties that border Medina County.
- Find the Counties that border Lower Merea County.

While some ontologists advocate merely using competency questions as an indication of the kind of question the RDFS ontology should answer (Noy and McGuinness, 2001), we would instead advise that there should be as many competency questions as necessary to cover all the queries you expect users of your data will want to ask. This follows the principles of test-driven development, and by the time you have a satisfactory list of competency questions, you will be well on your way to listing the classes and property names need for your RDFS ontology, which comes in stage 2 of the ontology design. The purpose, scope, and competency questions can be included in an `rdfs:comment` in the Linked Data to document the RDFS ontology.

As we have just mentioned, the second part of deciding what your data is about is to choose the class and property names. To do this, consider first the domain of interest. What are the most important concepts in the domain, and how are they linked together? For Merea Maps, the concepts will be things like Country, Region, District, County, City, and Parish. These are not hierarchically related (e.g., a City is not a kind of County), but they are related topologically, so a City will be in a County. We need to think a bit more about the topological spatial relations now, as "in" is quite a vague term. Merea Maps decides to use four topological relations:

- Spatially Contains: The interior of one region completely contains the interior of the other region. Their boundaries may or may not intersect.
- Spatially Contained In: The inverse of Spatially Contains. The interior of one region is completely contained in the interior of the other region. Their boundaries may or may not intersect.

- Borders: The boundaries of two regions intersect, but their interiors do not intersect.
- Spatially Equivalent: The two regions have the same boundary and interior.

Unlike in the ontology development of Chapter 10, we do not now start to link together the concepts and relationships in the RDFS ontology. Unless there are triples that are valid at the class level (i.e., true for every instance of the class), which is not often the case, this will not be necessary. The only exception to this is when we want to use the RDFS ontology to validate the data, that is, when we need to impose a requirement for every instance of the class on the data. If Merea Maps wanted to make sure that it was supplying quality data, such that every Parish included in its data was supplied with area information, it could link the class Parish to the relationship "has Area" with a datatype object. Think carefully before doing this, however, as it will mean that you will need to make sure you have the data available to instantiate the ontology. And having said that, it is now a good time to turn to the data and look at how it is structured.

7.4.2   STEP 2: LOOK AT THE CURRENT GI DATA

The reason for not looking at the data until the second step of the process is to avoid preconceived ideas about what we want the RDFS ontology to look like. Frequently, it is the case that the data is in a legacy format, and the current database or file format structure is a consequence of previous technological and implementation limitations rather than a requirement of the domain. This gives us a chance to start afresh with a new ontology. It is certainly possible to assign one URI to each row in the database table, spreadsheet, or comma-separated file, to create class names based on the table names and property names based on the column names, but they may not always be very meaningful to the outside world. Merea Maps has a dataset in a table, named in their database as "Regions," which is based on the structure of Table 7.1.

If alternatively Merea Maps was working with proprietary formats such as ESRI Shapefile or MapInfo Tab or MID/MIFF files for exchange of information, it would at this point look at the .dbf file (or similar) that stores the attribute information of the administrative region features.

There are a number of points to note from Table 7.1. First, not every field in the data record is needed. Some are only really internally relevant to Merea Maps. Other fields correspond to metadata, for example, edit date. Second, we do not need to stick to the field names that were previously chosen and perhaps shortened due to limitations on string length in the database technology. We can choose meaningful names for our classes and predicates. RDF has its own limitations on string names; for example, no spaces are allowed, so we can also use the rdfs:label to provide more human-friendly readable names. Third, we will reuse vocabularies and ontologies if possible. This not only saves us work but also makes linking to other datasets much easier. Merea Maps identified the WGS84 Basic Geo Vocabulary as useful for expressing latitude and longitude. Other potentially useful RDF vocabularies have already been outlined in Chapter 5.

TABLE 7.1

Table Describing the Data Structure of Merea Maps' Administrative Geographical Information

Field Name	Field Full Name	Example	Comments for the RDFS Ontology
F_ID	Feature ID	000123	A unique ID that can be used to generate the URI.
DEF_NAME	Definitive name	Ash Fleet Parish	"Definitive" is not an appropriate description in a decentralized Web world. Merea Maps will just use "Name" as the namespace will demonstrate that it is they who have coined the name.
DEG_LAT	Degrees of Latitude	54	Merea Maps decides to reuse the Basic Geo
MIN_LAT	Minutes of Latitude	2.7	(WGS84) vocabulary[a] that defines latitude and longitude and is used by many other Linked Data sets, so these fields will be combined.
DEG_LONG	Degrees of Longitude	0	These fields will be combined.
MIN_LONG	Minutes of Longitude	13.2	
X	Local X coordinate	462500	Outside scope
Y	Local Y coordinate	516500	Outside scope
Type	Type	Parish	This is actually the class to which the instance in this row belongs.
F_NAME	Filename	Medina County	This refers to the name of the file in which the polygon information is stored. In Merea Maps' case, the files are named according to the administrative area that contains the instance.
PID	Polygon ID	12345	A reference to the polygon in the file. Merea Maps *could* use this unique ID to form the URI; however, it identifies the spatial footprint, not the object itself, so it would be misleading to do so.
MAP_NO	Map number	647	Corresponds to Merea Maps' paper products. Outside scope.
TILE_NO	Tile number	839	Corresponds to Merea Maps' raster data product. Outside scope.
M_CODE	Merea Feature code	C	An internal code that is not needed.
E_DATE	Edit date	13-06-2001	When the data was collected and input to the system. This is useful but should be metadata, which we will discuss elsewhere.
UPDATE_CO	Update code	P	An internal code for Merea Maps indicating when the data should next be checked for accuracy. We could choose to include an "expiry date" on our data, but instead follow the convention of assuming it is in date unless deprecation is specifically noted. So, this is out of scope.

[a] http://www.w3.org/2003/01/geo/wgs84_pos#

7.4.3   STEP 3: SPECIFY YOUR RDFS ONTOLOGY

Merea Maps selects the following classes and properties for its Administrative Geography RDFS ontology:

```
Classes: Administrative Area, Country, Region, District, County, City,
and Parish
Properties: hasPlaceName (which is a specialization of foaf:name),
hasArea, spatiallyContains, spatiallyContainedIn, borders,
spatiallyEquivalent.
Imported properties: geo:lat, geo:long, rdf:comment, rdf:label
```

All the properties can be used by any class, but domain and range are not assigned to them to avoid overloading the vocabulary unnecessarily.

There are a number of tools available to assist with the development of an RDFS ontology. Most are also used to build the more complex OWL (Web Ontology Language) ontologies and hence are described in Chapter 9; however, you may also wish to look at Neologism,[4] a Web-based tool for creating and managing just RDFS vocabularies. Note that the RDFS ontology should also conform to Linked Data principles, namely, each class and relationship term should also be dereferenceable, so that Linked Data applications can locate their definitions.

A word of warning here: Do not try to add your own terminology to someone else's namespace, as you will have no control over terms in that namespace and will not be able to dereference your URIs or make sure that the namespace continues to exist in the long term. Instead, mint your own URIs and then use owl:sameAs to state their equivalence URIs in the namespace of your interest.

7.4.4   STEP 4: MINT YOUR URIs

In the previous chapter, we discussed at some length how to create URIs. We avoid including implementation details such as server names, file format, or port numbers and choose URIs of the format http://id.mereamaps.gov.me/ administrativeRegions/, using the 303 redirect ("slash") form. Although the general advice is to use unique keys from the domain if possible, rather than primary keys from your internal database, no obvious set of unique keys exists for Merean administrative regions (in the way that ISBN [International Standard Book Number] numbers are known keys for books). It may also be that another authority should really choose the identifiers for administrative areas, for example, the Merea Boundary Agency that assigns authority to the various administrative areas. But, to keep our example simple for now, MereaMaps uses the feature ID to generate its URI. Note that, as suggested previously, separate URIs should be created to describe:

a. The real-world object Ash Fleet Parish http://id.mereamaps.gov. me/administrativeRegions/000123

b. The HTML Web page with information about Ash Fleet Parish `http://page.mereamaps.gov.me/administrativeRegions/000123` and
c. The RDF/XML data representation of Ash Fleet Parish `http://data.mereamaps.gov.me/administrativeRegions/000123`

7.4.5 STEP 5: GENERATE YOUR LINKED DATA

Now that Merea Maps knows where its URIs are coming from, there are several ways of generating the actual RDF data. The choice depends both on what the input data is like and how the RDF is to be accessed on the Web. There are four main ways in which the data may originally be structured:

- Plaintext
- Structured data, such as comma-separated files, or spreadsheets
- Data stored in RDBs
- Data exposed via an Application Programming Interface (API)

RDF data can be published on the Web in one of the following ways:

- As a static RDF/XML file
- As RDFa embedded in HTML
- As a Linked Data view on an RDB or triple store

In addition to these three main publishing mechanisms, a publisher has the option of providing a data dump (that is, a zipped file containing all the RDF data, available for download) or a SPARQL endpoint (an interface for people or applications to query an RDF data store directly using the SPARQL query language). However, just providing these last two options for data access is not considered sufficient to be classified as "Linked Data."

We consider each of these input data structures and output publishing options in turn in Section 7.5. These options are equally applicable for GI and non-GI data; however, GI data may require some additional consideration and preprocessing. For example, queries may need to be run within the GIS or spatial database system itself to generate data about the areas of each polygon or topological relationships between the various administrative areas. This last point requires some design decisions from Merea Maps. Do they want to encode all possible triple relationships between every administrative area? In both directions? (For example, if "Area a borders Area b" is explicitly stated, should they also include the triple "Area b borders Area a"?) This is where a carefully designed scope and competency questions can come in: Merea Maps decides that the adjacency information ("borders") should be included in both directions (as it has competency questions in its list that can only be answered with such information), but the scope precludes the need to include topological relationships between entities of different classes. Similarly, containment relations are only explicitly stated between an administrative area and the areas that it immediately contains. There are a number of RDF features that should be avoided when creating Linked Data. These are reification, RDF containers and collections, and blank nodes. Reified statements are difficult to query with SPARQL (the RDF query language explained in Chapter 8). It is better to add metadata to the Web document that contains the triples instead. RDF collections and containers also cannot be queried in SPARQL, so if the relative ordering of items in a set is significant, add in multiple triples with the same subject URI and predicate and then add additional triples between the object URIs to explicitly describe the sequence information. Blank nodes cannot be linked to from outside the document in which they appear as their scope is limited to that document, and they pose a problem when data from different sources is merged as they cannot be referenced by a URI. So, the recommendation is to name every resource in the dataset with an explicit URI.

7.5   LINKED DATA GENERATION

7.5.1   PLAINTEXT DATA SOURCES

Although this is less common in a purely GI environment, there are many scenarios for which the contents of text documents need to be converted to Linked Data, for example, news stories, patents, or historical records, and these, like many information sources, are likely to include some references to location as well. A tool such as Open Calais[5] or Ontos Miner[6] (which can be applied to a number of languages other than English) can identify the "entities"—the main people, organizations, places, objects, and events in the text—using various natural language-processing and machine-learning techniques and assign URIs to them. However, a word of warning here: These tools get it wrong a lot of the time; typically, precision rates are only around 80%, so they should not be used without manual verification. Usually, the resulting RDF is embedded as RDFa metadata alongside the text as it is published on the Web (as explained in Section 7.5.6), making the text documents more easily discoverable and enabling faceted browsing. However, it is equally possible for the RDF extracted from the plaintext to be stored in a triple store that is published to the Web or simply published as a static RDF/XML file.

7.5.2   STRUCTURED DATA SOURCES

In contrast to plaintext documents, GI data is more usually accessible in some structured format, for example, CSV (comma-separated values), XML, or even an Excel spreadsheet. If the data from the GIS can be output as comma-separated files, code can be written in scripting languages such as Perl to convert to RDF/XML structure. If the GI is in an XML-based format, XSLT transformations are possible instead. There are several "RDF-izer" tools available to assist in this process. The tools usually convert the original structured data format to a static RDF file or load the RDF data into a triple store. These include Excel/CSV converters from Cambridge Semantics,[7] Topbraid,[8] and XLWrap.[9] A more comprehensive list of RDF conversion tools has been collected by the W3C and is available on its wiki.[10]

7.5.3 RELATIONAL DATABASE SOURCES

Merea Maps stores its data in a GIS that is implemented as a software layer on top of an RDB, an architecture that is common to many GIS systems. Merea Maps has the option of outputting the data from its GIS into CSV or XML and converting it to RDF using one of the RDF-izer tools discussed. However, since it is continuing to use its GIS as the mainstay of its business, and the data is continuously updated as new houses are built, the landscape changes, and so on, it would be better for it to publish a Linked Data view of its RDB. This has the advantage that there is no disruption to the existing business uses of the database or GIS, and the Linked Data remains in sync with the latest version of Merea Maps' GI data.

The simplistic approach to an RDB to RDF mapping (Berners-Lee, 1998b) is

- An RDB table record is an RDF node.
- The column name of an RDB table is an RDF predicate.
- An RDB table cell is a value.

Each table represents an RDFS class, and foreign key relationships between tables are made explicit by including a triple relating one RDFS class to another. A simplistic example of an RDB table is shown in Table 7.2. The data in Table 7.2 can be converted to the following RDF (shown in Turtle format here):

```
@prefix rdf: <http://www.w3.org/1999/02/22-rdf-syntax-ns#>.
@prefix rdfs: <http://www.w3.org/2000/01/rdf-schema#>.
@prefix mm_address: <http://id.mereamaps.gov.me/addresses>.
mm_address:0001 House_Number "39";
    Street "Troglodyte Road".
mm_address:0002 House_Number "45";
    Street "Isis Way".
```

As we can see, since the Geo IDs are unique identifiers (and primary keys of the table), they can be reused within URIs. Each row of the table represents one resource, and each column heading is used as a property for a triple. The geometry data (a Binary Large Object or BLOB[11] type) can be represented as the ogc:Polygon type, introduced in Section 6.4. The main problem that arises, however, is that the table structure of Merea Maps' original data from Table 7.1 does not match the RDFS ontology that has been developed in Section 7.4.3 to more accurately represent the domain.

TABLE 7.2

Simplistic Example of a Relational Database Table

Geo ID	House_Number	Street	Geometry
00001	39	Troglodyte Road	[BLOB]
0002	45	Isis Way	[BLOB]

To resolve this, a BLOB is a package of data with internal structure defined by the originator and normally requiring custom code to interpret to specify a mapping from the RDB structure to the RDF graph. There are several tools that facilitate the publication of a Linked Data view of the RDB by allowing the data publisher to specify these mappings. These include Virtuoso[12] from Open Link, Triplify (Auer et al., 2009), R2O (Barrasa and Gómez-Pérez, 2006), and D2R Server.[13] Merea Maps uses one of these tools to automatically generate the "simplistic" version of an RDB-to-RDF mapping[14] as it is a good starting point, and then it customizes the mapping to generate the RDF data corresponding to the RDFS ontology it actually wants. This includes removing some of the mappings for columns they do not want to keep (such as the Merea Feature Code) or writing more complex mappings for an amalgamation of columns (such as Degrees and Minutes of Longitude).

Currently, each RDB-to-RDF tool has its own proprietary way of mapping from the relational schema to the RDF view. Some allow custom mappings, while others are limited to simplistic mappings only.

7.5.3.1 Virtuoso

The Virtuoso RDF Views software from Open Link creates "quad map patterns" using the Virtuoso Meta Schema language to define mappings from a set of RDB columns to triples in a specific graph (that is, a "quad" of graph, subject, predicate, object). Virtuoso's SPARQL-to-SQL translator recognizes triple patterns that refer to graphs that have an RDB representation and translates these into SQL accordingly. The main purpose of this is to map SPARQL queries onto existing RDBs, but it also enables integration with a triple store. Virtuoso can process a query for which some triple patterns can be matched against local or remote relational data and some matched against locally stored RDF triples. Virtuoso tries to make the right query compilation decisions based on knowledge of the data and its location, which is especially important when mixing triples and relational data or when dealing with relational data distributed across many external databases. Virtuoso covers the whole relational model, including multipart keys and so on.

7.5.3.2 Triplify

Another tool, Triplify,[15] is a software plug-in to Web applications that maps HTTP GET requests onto SQL queries and transforms the results into RDF triples. These triples can be published on the Web in various serializations: RDF/XML, JSON (Java Script Object Notation), or Linked Data (i.e., resources with resolvable URIs that are linked together on the Web). The software does not support SPARQL queries, but does enable the publishing of update logs, to facilitate incremental crawling of Linked Data sources, as well as an extension to specify provenance. Also of note is that Triplify has been used to publish data from the OpenStreetMap project as RDF[16]: a very large (greater than 220 million triples) dataset. However, the drawback of Triplify's approach is that semantic meaning is encoded in the query itself and is not easily accessible, or indeed decipherable, outside the Triplify system. This problem is demonstrated in a query[17] used to enable the LinkedGeoData "near" REST service that returns RDF for nearby point locations:

```
$triplify['queries'] = array(
    '/^near\/(-?[0-9\.]+),(-?[0-9\.]+)\/([0-9]+)\/?$/' =>'
SELECT CONCAT("base:",n.type,"/",n.id,"#id")
AS id, CONCAT("vocabulary:",n.type)
AS "rdf:type", '.$latlon.',
    rv.label AS "t:unc", REPLACE(rk.label,":","%25"), '.$distance.'
FROM elements n
    INNER JOIN tags t USING(type,id)
    INNER JOIN resources rk ON(rk.id = t.k)
    INNER JOIN resources rv ON(rv.id = t.v)
WHERE '.$box.'
HAVING distance < $3 LIMIT 1000',
);
```

To find places near (within a certain radius) to a certain latitude/longitude point, one can type http://linkedgeodata.org/triplify/near/*latitude,longitude/radius* into any Linked Data-enabled browser. However, one still needs to know (by reading the documentation) how to structure the URL (Uniform Resource Locator) to access the Web service and understand (by understanding the SQL query and regular expressions, as discussed) how the results have been calculated. It is not yet a transparent, explicit, RDF-based approach.

7.5.3.3 R2O

R2O is another language for expressing mappings between RDBs and an RDFS ontology, this one based on XML. It works alongside ODEMapster, a processor that uses the R2O document to create an RDF dump of the whole database or, when in query-driven mode, translates queries expressed in the ODEMQL query language (which is specifically designed for ODEMapster) on the fly to SQL. It has been used extensively by the GeoLinkedData project[18] (not to be confused with the Linked GeoData RDF version of GeoNames), which is tasked with adding Spanish geospatial data to the Linked Data Web. As well as requiring queries to be written in a proprietary language, rather than SPARQL, ODEMapster cannot in itself serve Linked Data. Instead, the GeoLinkedData project has to load the generated static RDF into Virtuoso's triple store to publish on the Web. Hence, it is less suitable for Merea Maps' direct use.

7.5.3.4 D2R Server

Another tool that Merea Maps could use to expose a Linked Data view of its RDB is D2R Server. D2R Server offers a Linked Data interface (also known as a "dereferencing interface") that makes RDF descriptions of resources available via HTTP. Using a Linked Data-enabled browser such as Tabulator or Disco (see Section 7.10), the RDF description can be retrieved over the Web just by entering the URI into the browser's address box. D2R Server also offers a normal HTML Web interface and a SPARQL endpoint. HTTP requests are rewritten into SQL queries, using the mapping, so as with Virtuoso, the RDF data can be explored on the fly, without the need to replicate the data in a dedicated RDF triple store. D2R Server uses its own proprietary mapping language called D2RQ, which describes the relationships between the RDFS ontology and the relational data model. Let us look at the Merea Maps example expressed in the D2RQ Mapping language. As you can see, it is in Turtle format, using some specific d2rq vocabulary items. Lines 1 and 2 specify the namespaces for RDF and RDFS ontology, with which we are familiar, while line 3 specifies the location of this D2RQ vocabulary. Line 4 gives a prefix for the mapping file's namespace, although this does not much matter as it will not appear in the final RDF. The prefix in line 5 is for the jdbc namespace, which the D2R Server code uses to link to the database, and line 6 denotes Merea Maps' administrative regions' RDFS namespace. Lines 8 to 12 specify an instance of the D2RQ "Database" class and includes the instructions for how to connect to the database. In this case, Merea Maps has an Oracle database, but D2RQ works with MySQL, PostgreSQL, and Microsoft SQL Server as well. The strings in jdbcDriver and jdbcDNS (the JDBC URL connection for the database) will depend on which type of database you use. An instance of d2rq:ClassMap is created in lines 14–18 and includes information about where the data is stored (line 15) and the structure of the resources' URIs (line 16). In this case, it will be of the form "http://data.mereamaps.gov.me/administrativeRegions/" concatenated with the contents of the Regions table "F_ID" column. Line 17 makes the RDF resource an instance of the RDFS class admin:Parish, and line 18 indicates that only those Regions table rows that meet the condition where the DESC column contains the string "Parish" should be returned. (This corresponds to an SQL FILTER statement.)

```
1. @prefix rdf: <http://www.w3.org/1999/02/22-rdf-syntax-ns#>.
2. @prefix rdfs: <http://www.w3.org/2000/01/rdf-schema#>.
3. @prefix d2rq: <http://www.wiwiss.fu-berlin.de/suhl/bizer/D2RQ/0.1#>.
4. @prefix map: <file:/MereaMaps/adminRegions-mapping.n3#>.
5. prefix jdbc: <http://d2rq.org/terms/jdbc/>.
6. @prefix admin: <http://data.mereamaps.gov.me/administrativeRegions/>.
7.
8. map:database a d2rq:Database;
9.     d2rq:jdbcDriver "oracle.jdbc.driver.OracleDriver";
10.    d2rq:jdbcDSN "jdbc:oracle:thin:@merea-server:1521:merea-db";
11.    d2rq:username "root" ;
12.    d2rq:password "password".
13.
14. map:parishMap a d2rq:ClassMap;
15.    d2rq:dataStorage map:database;
16.    d2rq:uriPattern
"http://data.mereamaps.gov.me/administrativeRegions/@@Regions.F_ID@@";
17.    d2rq:class admin:Parish ;
18.    d2rq:condition "Regions.DESC = 'Parish'".
19.
20. map:parishMap_hasPlaceName a d2rq:PropertyBridge;
21.    d2rq:belongsToClassMap map:parishMap;
22.    d2rq:property admin:hasPlaceName;
23.    d2rq:column "Regions.DEF_NAME".
24.
25. map:parishMap containedIn a d2rq:PropertyBridge;
26.    d2rq:belongsToClassMap map:parishMap;
```

```
27.     d2rq:property admin:spatiallyContainedIn;
28.     d2rq:column "Regions.F_NAME".
29.
30.map:parishMap latitude a d2rq:PropertyBridge;
31.     d2rq:belongsToClassMap map:parishMap;
32.     d2rq:property geo:lat;
33.     d2rq:sqlExpression "SELECT Regions.DEG_LAT + Regions.MIN_LAT/60".
34.
35.map:parishMap longitude a d2rq:PropertyBridge;
36.     d2rq:belongsToClassMap map:parishMap;
37.     d2rq:property geo:long;
38.     d2rq:sqlExpression "SELECT Regions.DEG_LONG + Regions.MIN_
LONG/60".
```

Lines 20 to 23 create the mapping for the property admin:hasPlaceName. Line 21 states that it belongs to the d2rq:ClassMap parishMap. So, triples will be created that have subject URIs of the form in line 16, the predicates will be admin:hasPlaceName, and the objects will be literal values from the database table Regions, column "DEF_NAME." A similar pattern is followed in lines 25 to 28 to create the mapping for admin:spatiallyContains. Lines 30 to 33 and 35 to 38 specify the mappings for the geo:lat and geo:long properties, respectively. The value of the triple's object is the result of the SQL query stated in d2rq:sqlExpression (lines 33 and 38), namely, the sum of the values in the DEG_LAT and MIN_LAT/60 columns (or DEG_LONG and MIN_LONG/60). In this way, the conversion from degrees and minutes to decimal latitude and longitude can be specified in the mapping itself. However, carrying out this kind of calculation on the fly tends to degrade performance, and it might be better to create a new database table, or at least View on the database, that had precalculated the latitude and longitude in decimal format.

7.5.3.5   R2RML

A standard language, R2RML, to express mappings from RDBs to RDF is currently being developed by the W3C RDB2RDF Working Group (Das, Sundara, and Cyganiak, 2011). Since the mappings are themselves RDF graphs, the knowledge encoded in the mappings are made explicit and more easily understandable than information encoded in any of the proprietary mapping languages. Hence, this initiative will go some way to resolving the problem of knowledge being hidden in the mapping itself. It will also enable the mappings to be reused should the underlying RDB be migrated to a different vendor's and simplify programming applications that access multiple data sources. Thus, currently there is no "perfect" solution to recommend to Merea Maps, although as we have seen, Virtuoso and D2R Server offer two useful options, the former commercially supported and the latter open source.

7.5.4   DATA SOURCES WITH APIS

So far, we have discussed data sources that are text documents, structured data such as CSV files or spreadsheets, and RDB sources. But, what happens when the data is hidden behind a proprietary Web API, such as the Flickr, Amazon, Twitter, or Facebook APIs? In this case, you need to write your own conversion script or "wrapper," which needs to take into account the individual API's query and retrieval interface and its particular way of returning results, which could be in ATOM, JSON, or XML formats, to name the most common. The wrapper needs to assign a URI to the data resource provided by the API, such as a tweet or a photo; then when an HTTP request for the URI is received, the wrapper rewrites the request into the format that the underlying API requires. The API returns the data in some format, such as XML or JSON, and the wrapper then transforms this into RDF/XML.

A few wrappers are already in existence, for example, ShreddedTweet[19] (although this does not yet serve Linked Data, it does convert to RDF via RDFa); Flickr2rdf[20]; and the RDF Book Mashup,[21] and the process will become easier as more reusable open source modules (such as the RDF API for PHP[22]) are written. When the HTTP request for a book URI is received, the RDF Book Mashup script pulls out the ISBN number from the URI and uses this to query the GoogleBase and Amazon APIs. The APIs return XML, and this is serialized into RDF/XML. Similarly, Flick2rdf queries the Flickr API and puts the resulting information on photo metadata and tags into RDF triples. Perhaps one day the API owners will themselves provide RDF conversion wrappers as a matter of course.

7.5.5 PUBLISHING STATIC **RDF/XML** FILES

Publishing static RDF/XML files is most suitable when you wish to publish only a small amount of data, which is unlikely to change much, as this method involves the least effort on the part of the publisher. If the dataset is larger, it is better to split it into several files to avoid delay when a browser is forced to load and parse a large RDF file. The RDF/XML file is created once, either manually, such as when an RDFS vocabulary is created, or when a software tool exports RDF/XML from a triple store. The file can be published on the Web using a standard Web server, configured to return the correct MIME type (`application/rdf+xml`), so that Linked Data applications will recognize the data as RDF. Clearly, this method is less useful for datasets that frequently change as the RDF/XML would need to be re-created and republished every time.

7.5.6 PUBLISHING AS **RDFA** EMBEDDED IN **HTML**

Another way to publish data is to embed it as RDFa within an HTML Web page (see Section 5.5.4 for details of RDFa). The advantage of this is that only one document needs to be updated if the information changes and is particularly suited for content that is already part of an existing content management system, where the templates for creating the HTML can be extended to include RDFa. Some content management systems, such as Drupal,[23] are now providing support for publishing RDFa. The W3C's RDFa Distiller and Parser tool[24] can be used to convert RDFa embedded in HTML into RDF serialized in Turtle or RDF/XML, if the data needs to be processed independently of the HTML page.

This form of publishing would be suitable if Merea Maps wanted to, say, include its address as RDF data on its organization's Web site or embed data about its catalogue of mapping products, that is, information that is related to the HTML content and hence can help with faceted browsing or semantic search. There is less of a compelling use case for encoding GI data such as that in Table 7.1 in RDFa as it is not directly related to any human-readable information on an HTML Web page and can only really be used and processed if it is converted from RDFa to RDF.

7.5.7 PUBLISHING A LINKED DATA VIEW ON A RELATIONAL DATABASE OR TRIPLE STORE

Since Merea Maps has a large volume of data that changes frequently and is not directly related to particular HTML Web pages, the most obvious approach for its RDF publication would be as a Linked Data view on an RDB or triple store. If it wants to stick to just one workflow involving its existing GIS, a Linked Data view of its RDB can be published using a tool such as Virtuoso or D2R Server, as described in Section 7.5.3. If, however, Merea Maps wanted to store its RDF data in a manner more suited to the RDF graph structure, it would be well advised to use a "triple store," a non-RDB optimized for storing Linked Data. Particular examples of these are discussed in Section 7.10 on software tools. Many triple stores provide a Linked Data interface to make RDF descriptions of stored resources available via HTTP; however, for those that do not include a Linked Data interface as standard (that is, they store the data using URIs that are not dereferenceable), there are tools such as Pubby[25] and Djubby[26] that sit in front of the triple store's SPARQL endpoint to serve Linked Data. When the HTTP request comes in as a URI for a Linked Data resource, Pubby rewrites the request into a SPARQL DESCRIBE query (see Chapter 8) by mapping the dereferenceable Linked Data URI onto the original triple store's URI, and the query is then put to the triple store. Pubby also handles 303 redirects (see Section 7.3.1) and negotiates between the HTML, RDF/XML, and Turtle descriptions of the resource. Djubby operates in a similar way but is written in Python rather than Java and can be used with the Django Web framework.

7.6 DESCRIBING THE LINKED DATASET

Now that Merea Maps has been able to create and serve up its Linked Data, it needs to consider how people can find this Linked Data and know what it is about. To assist data discovery, any related HTML documents should have a <link> tag inserted into the header to point to the RDF file. This helps make the RDF visible to Web crawlers and Linked Data browsers and is known as the Autodiscovery pattern (Dodds and Davis, 2012):

```
<link rel = "alternate" type = "application/rdf+xml" title =
"Topographic Object Data" href = "http://mereamaps.gov.me/topo.rdf">
```

Merea Maps provides metadata alongside its data to assist third parties' understanding of what is offered. The purpose, scope, and competency questions, provenance, licensing, and currency (when the dataset was last updated) are all important elements of the metadata that should be provided with the Linked Data set to encourage not only reuse but also *accurate* reuse. There are currently two methods available for publishing descriptions of a Linked Data set: Semantic Sitemaps (Cyganiak, Delbru, and Tummarello, 2007) and the Vocabulary of Interlinked Datasets (VoID) (Alexander et al., 2011).

7.6.1 SEMANTIC SITEMAPS

The Semantic Sitemaps protocol provides information to Web crawlers about the pages available in a Web site. An XML document sitemap.xml is stored in the Web site's root directory and encodes information about the URL, the site's location, when it was last modified, and how often the site changes to help a search engine optimize how frequently it needs to crawl the site. A Semantic Sitemap is an extension to the standard sitemap idea, adding information about the dataset such as a URI and label for the whole dataset, as well as pointers to its SPARQL endpoint. It also can include details of where the whole dataset can be downloaded from or information about the shape of the RDF graph. The semantic search engine can go to the dataset's URI to find an additional RDF description about the dataset. The sitemap's "slicing" attribute specifies which shape graph is returned when a URI is dereferenced. This is useful for the crawler to know in advance so that it knows how much detail to expect. There are various ways to slice an RDF graph, depending on how much detail about the URI should be returned. One common shape is the concise bounded description (Stickler, 2005), which is a unit of specific knowledge about a resource that can be used or exchanged by different software agents on the Semantic Web. Given a particular node (the starting node) in a particular RDF graph (the source graph), the concise bounded description subgraph is

- All statements in the source graph where the subject of the statement is the starting node.
- For each of these statements that have a blank node as an object, recursively add in all further statements in the source graph that have the blank node as a subject.
- If there are any reifications in the source graph, recursively include all concise bounded descriptions beginning from the rdf:Statement node of each reification.

The result of these steps is a subgraph where the object nodes are URI references, literal values, or blank nodes that are not the subject of any statement in the graph. This is also known as the "bnode closure" of a resource.

Since it is preferable in RDF not to worry about the direction of a property, an alternative shape, the symmetric concise bounded description, has also been proposed, which includes all statements on both the outbound and inbound arc paths of the starting node.

7.6.2 VOCABULARY OF INTERLINKED DATASETS

While the Semantic Sitemap follows the standard Sitemap protocol and is encoded in XML, VoID[27] offers more self-consistency as it is itself encoded in

RDF. It includes some of the same information as the Semantic Sitemap (such as the data dump and SPARQL endpoint) but also describes the vocabulary used in the dataset and its outgoing and incoming links, as well as specifying any subsets of the dataset.

There are two main ways of encoding metadata in RDF: reification and Named Graphs. As we saw in Chapter 5, the technique of reification specifying an `rdf:Statement` with a subject, predicate, and object allows other information, namely, metadata, to be added about the `rdf:Statement`. There are several drawbacks to this approach, however. First, it makes querying difficult as the triple graph model has been broken. Second, it buries metadata within the data itself, potentially making the volume of data balloon and forcing a user to deal with both data and metadata at the same time. VoID takes an alternative approach, to assign a URI to the graph itself, and then metadata about the graph can be also assigned to this URI. The graph is then known as a Named Graph as it has been given a URI. This approach maintains the RDF triple structure throughout both the data and metadata, and the metadata can conveniently be kept separate from the main data. The downside is that a system of unique identifiers must be maintained for every named graph in the dataset, which becomes more of a management issue as the granularity of the metadata becomes finer. For example, while it is straightforward enough to create one URI for the whole dataset and assign metadata to the whole dataset, if we wish to add different metadata to each triple (which might be particularly useful if the triples have different provenance), we will need to provide a separate URI to each "Named Triple."

The VoID vocabulary defines the class `void:Dataset`, an instance of which is an RDF resource that represents the whole dataset, so that statements can be made about the whole dataset's triples. VoID also describes linksets, collections of RDF links between two datasets. A link is an RDF triple whose subject is in one dataset and object in another. The `void:Linkset` class is a subclass of `void:Dataset`. The following shows an example of VoID metadata for Merea Maps' administrative geography RDF dataset:

```
1.  <?xml version = "1.0"?>
2.    <rdf:RDF xmlns:foaf = http://xmlns.com/foaf/0.1/
3.         xmlns:void = http://rdfs.org/ns/void/
4.         xmlns:rdf = http://www.w3.org/1999/02/22-rdf-syntax-ns#
5.         xmlns:dcterms = "http://purl.org/dc/terms/">
6.     <void:Dataset rdf:about = "http://data.mereamaps.gov.me">
7.       <foaf:homepage rdf:resource = "http://page.mereamaps.gov.me"/>
8.       <dcterms:title>Administrative Geography</dcterms:title>
9.       <dcterms:description>An example Merean dataset in
10.        RDF.</dcterms:description>
11.       <dcterms:publisher rdf:resource = "http://mereamaps.gov.me"/>
12.       <dcterms:license
13.         rdf:resource = "http://www.opendatacommons.org/licenses/
pddl/1.0/"/>
14.      <void:sparqlEndpoint rdf:resource = "http://mereamaps.gov.me/
sparql"/>
```

```
15.    <void:uriLookupEndpoint rdf:resource = "http://lookup.mereamaps.
gov.me/"/>
16.    <void:dataDump rdf:resource = "http://mereamaps.gov.me/dump.rdf"/>
17.    <void:vocabulary rdf:resource = "http://purl.org/dc/terms/"/>
18.    <void:vocabulary rdf:resource = "http://mereamaps.gov.me/
admingeo"/>
19.    <void:exampleResource rdf:resource = "http://data.mereamaps.
gov.me/000123"/>
20.    </void:Dataset>
21.</rdf:RDF>
```

There are four types of metadata specified by VoID: general metadata using the Dublin Core vocabulary; metadata describing how to access the data; structural metadata, which can be used for query and integration; and finally metadata denoting links between the dataset in question and third-party ones on the Linked Data Web. The general metadata includes information about any home page associated with the dataset (line 7), contact e-mail address of the author or publisher can be supplied, and Dublin Core metadata tags can be used to supply information such as licensing or subject of the dataset (lines 9 to 13). The `void:feature` can be used to specify which RDF serialization format is used, for example, RDF/XML or Turtle. Access metadata is used to describe methods of accessing the RDF triples in the dataset (note that the dataset itself is not part of the VoID description). For example, the access metadata can include how RDF data can be accessed via a SPARQL endpoint (line 14), a URI lookup (line 15), as a dump of RDF data (line 16), or if the data publisher offers a search facility into its data, this can also be described using a standard OpenSearch[28] description document via the term `void:openSearchDescription`.

The structural metadata describes the internal structure of the dataset and provides additional information about the schema, which can be useful to explore or query the dataset. For example, the vocabularies used by the dataset (using `void:vocabulary` as in lines 17 and 18), the size of the dataset (using `void:entities, void:triples, void:classes,` or `void:properties` to express the number of entities, triples, classes, or properties, respectively), or examples of resources that are in the dataset. The void term `void:uriRegexPattern` specifies the regular expression that matches the URIs of the resources in the dataset. Another description that can be used is `void:uriSpace,` which specifies the namespace to which all the resources belong. The structural metadata can in addition include information about how the dataset can be partitioned into subsets using the `void:subset` property. Each part of the dataset is itself a `void:Dataset,` which can be independently annotated with any VoID annotation. A dataset may need to be partitioned if different parts originate from different sources, the parts have different publication dates, the parts can only be accessed via different methods or at different locations (e.g., via different SPARQL endpoints or different RDF dumps), or perhaps the most obvious reason for dividing a dataset, the parts are concerned with different topics (i.e., they have different `dcterms:subject` values). Alternatively, dataset segmentation can be carried out on a class or property basis. Thus, one class-based partition (using `void:classPartition`) would include all the data that describes instances of a particular class, while a property-based partition (that uses `void:propertyPartition`) contains all the triples that use that property as the predicate.

The final type of metadata that can be described using a VoID vocabulary describes the links between two sets of RDF triples. Every set of links (a `void:linkset`) must have exactly two distinct `void:targets` (a subject and an object dataset). For example, if Merea Maps wanted to specify a linkset connecting the Merea Maps administrative regions dataset to the Merean Mail postcode dataset, it could specify a linkset as follows:

```
:MereaMaps_MereanMail a void:Linkset;
void:target http://data.mereamaps.gov.me;
void:target http://postcodes.mm.gov.me;
```

There are several tools for helping you generate a VoID description, including ve2, the VoID editor,[29] and voidGen.[30] VoID can also be produced within the OpenLink Virtuoso triple store,[31] which is discussed further in this chapter. The VoID browser is an easy way to view and query VoID descriptions of datasets. There are several different options for publishing a VoID description. One is to embed the VoID as RDFa markup into an HTML page about the dataset, using a local "Hash URI" for the dataset, for example, http://mereamaps.gov.me/#AdministrativeGeography. An alternative option is to use a "Slash URI" and along with the 303 redirect mechanism explained previously in this chapter, to serve the VoID in the same way as the main RDF dataset. The final option is to encode the VoID description in the RDF format Turtle. The file is, by convention, called void.ttl, and should be placed in the root directory of the site with a local "Hash URI." The main RDF dataset will then have a URI of the form http://mereamaps.gov.me/void.ttl#AdministrativeGeography.

7.7   PROVENANCE

To facilitate future reuse, it is important to specify where the data has come from: its *provenance*. Provenance is defined as "an explicit representation of the origin of data"[32] and can include information about how a dataset was produced or what facts were relied on for a decision. On one level, a dereferenceable URI does this implicitly by providing information directly from the owner of the URI. However, as Merea Maps takes data from the Merean Post Office to create its dataset of addresses, it also wants to be able to specify which fragments of its address data came from this third party. This provenance metadata can be represented as RDF triples, where the object of the triple specifies the document in which the original data sits. As mentioned, the Dublin Core vocabulary can be used to indicate who created and published the data and when using the predicates `dc:creator`, `dc:publisher`, and `dc:date`. If the objects of these creator and publisher triples are the URIs of the relevant people or organizations, rather than just their names as strings, further information about the provenance of the data can be obtained by querying the dataset for any triples with the creator or publisher URI as the subject.

Other suggestions have been put forward for encoding provenance, for example, the Open Provenance Model (Moreau et al., 2011), an OWL ontology[33] that allows additional descriptions of provenance based on Agents (authors, publishers, etc.), Processes (e.g., reslicing of data), and Artifacts (e.g., an RDF graph that was generated by a process). At the time of writing, there were several other vocabularies that can be used to describe provenance, including the Changeset Vocabulary,[34] Provenance Vocabulary,[35] and Semantic Web Publishing Vocabulary.[36] A W3C Provenance Interchange Working Group[37] is under way, tasked with providing mappings between these various provenance formats. As the technology is in a state of flux, with no clear *de facto* standard, we just recommend that the provenance of your GI Linked Data is specified using one of these vocabularies. For those hoping to reuse your data, it is useful to include descriptions of who has written and published your GI Linked Data and any limitations on the accuracy or frequency of your surveys or other data-gathering techniques.

7.8 AUTHENTICATION AND TRUST

A word now on the various other aspects of data quality assessment, namely, authentication and trust. Authentication contributes to the establishment of trust and includes mechanisms such as verifying a URI, controlling access to a resource, or using digital signatures, while trust is more of a social concept and remains harder to mechanize.

The Named Graphs API for Jena[38] (NG4J) is one software library that can be used to produce digital signatures for Linked Data and contribute to the authentication process as it can be particularly helpful in verifying that the provenance metadata does indeed belong to the Linked Dataset itself. The method NG4J uses to sign and store the digital signature of a Named Graph is first to find its canonical representation, that is, a representation that specifies which nodes of the graph are adjacent to which other nodes. Second, a digest of the canonical graph is calculated using any common secure hash function (for example, SHA-1). The digest is represented as its own named graph, which is called the Warrant Graph. In turn, the canonical representation of the Warrant Graph is taken and signed with the data publisher's private key using a standard signature algorithm like DSA or RSA. This signature is added to the Warrant Graph, and the signed Warrant Graph can then be published. To check whether a digital signature of a named graph is valid, the NG4J software will carry out the following verification process: First, the digital signature is extracted from the warrant graph of the named graph, along with the public key of the information publisher. The public key is used to verify the signature of the Warrant Graph, that is, to check that the signature does indeed belong to the information publisher. Second, the canonical representation of the named graph is found and a digest created using the SHA-1 hash function. This digest is compared against the digest in the warrant graph, and if they are the same, then the named graph has a valid signature.

While provenance provides the input information to a trust measurement algorithm, the degree of trust itself is the result of the question: Is this data good enough to use? and is often based in part on who else thinks the data is good enough to use.

Trust measurement is a very open area of research, and there are not yet any satisfactory automatic and general solutions to the problem. The best advice we can give at present, when evaluating a third-party Linked Data set for potential reuse, is to manually consider the provenance metadata and use your best judgment. We discuss more about how to assess the provenance of other people's data in Chapter 8 on Linking Data.

7.9 LICENSING LINKED DATA

According to Bizer, Jentzsch, and Cyganiak (2011), 85% of Linked Data does not include any licensing information. Specifying the terms under which the data can be reused helps encourage its reuse as it reduces uncertainty about what is allowed. Hence, it is good practice to include information about the licensing restrictions on any Linked Data that you publish within its metadata or, if there are no restrictions, as encouraged in the Linked Open Data movement, to include a waiver statement. You should specify your license in RDF (and there are many licenses already available that you can reuse, discussed in a separate section) and then indicate where your license file is to be found by including a triple in your dataset with the predicate `dcterms:license`.

7.9.1 OPEN LINKED DATA

The definition of *open* content or data is that "anyone is free to use, reuse, and redistribute it—subject only, at most, to the requirement to attribute and share-alike."[39] Tim Berners-Lee has a five-star rating system for the openness of Linked Data, and it is interesting to note that he places more emphasis on the openness of the data than on the ease of linking. The rating system (Berners-Lee, 2006) is as follows:

★   Available on the web (whatever format) *but with an open licence, to be Open Data*
★★   Available as machine-readable structured data (e.g., excel instead of image scan of a table)
★★★   as (2) plus non-proprietary format (e.g., CSV instead of excel)
★★★★   All the above, plus: Use open standards from W3C (RDF and SPARQL) to identify things, so that people can point at your stuff
★★★★★ All the above, plus: Link your data to other people's data to provide context

7.9.2 LINKED DATA LICENSES

The Creative Commons[40] is a frequently used licensing framework for the World Wide Web that offers several licensing options, including allowing or preventing commercial use, modification, and attribution of the work. A license can be encoded in RDF using the Creative Commons Rights Expression Language, which is specified under the namespace <http://creativecommons.org/ns#> and then referred to in the Linked Data set using the triple (shown in Turtle format)[41]:

```
@prefix cc: <http://creativecommons.org/ns#>.
cc:license <http://creativecommons.org/licenses/by-sa/3.0/rdf>.
```

This particular license (CC-BY-SA) meets the Open Linked Data principles and, expressed in Turtle format, is as follows:

```
1   @prefix rdf: <http://www.w3.org/1999/02/22-rdf-syntax-ns#>.
2   @prefix dc: <http://purl.org/dc/elements/1.1/>.
3   @prefix dcq: <http://purl.org/dc/terms/>.
4   @prefix cc: <http://creativecommons.org/ns#>.
5
6   <http://creativecommons.org/licenses/by-sa/3.0/>
7       rdf:type cc:License ;
8
9       cc:requires
10        cc:Notice,
11        cc:ShareAlike,
12        cc:Attribution ;
13        cc:legalcode <http://creativecommons.org/licenses/by-
14          sa/3.0/legalcode"/>;
15        dc:creator <http://creativecommons.org"/> ;
16        cc:permits
17          cc:Distribution,
18          cc:Reproduction,
19          cc:DerivativeWorks ;
20        dc:identifier "by-sa"
21        cc:licenseClass <http://creativecommons.org/license/">;
22        dc:title "Attribution-ShareAlike 3.0 Unported";
23        dcq:hasVersion "3.0".
```

The use of the CC-BY-SA license for data, however, is sometimes discouraged as it was not specifically designed for data but for any type of content. Some more suitable open licenses, designed for data, include

- http://www.opendatacommons.org/licenses/pddl/1.0/ (public domain for data/databases)
- http://www.opendatacommons.org/licenses/odbl/1.0/ (attribution for data/databases)
- http://www.opendatacommons.org/licenses/by/1.0/ (attribution and Share-Alike for data/databases)
- http://creativecommons.org/publicdomain/zero/1.0/ (Creative Commons Public Domain Waiver; see Section 7.8.3)
- http://reference.data.gov.uk/id/open-government-licence (an open license used by the U.K. government)

However, if for commercial reasons you need to place licensing restrictions on your data, the process for adding a license remains the same: create a license in RDF and link your dataset to it using the dcterms:license property.

7.9.3 LINKED DATA WAIVERS

A waiver is the opposite of a license: It allows data creators to unequivocally declare that they do not want to assert legal rights to their data. To do this, the data creators can simply state that their license is the Creative Commons Public Domain Waiver[42] using the dc:license predicate. However, even though the publisher has waived its rights, it may want to make sure that no one else can curtail the use of its data. One vocabulary that can assist in doing this is the Waiver Vocabulary,[43] which includes a predicate wv:norms, allowing the data owner to indicate the expectations it has about how its data will be used. This community norm is a nonbinding condition of use that the publisher encourages users to abide by. One commonly used example of this is the Open Data Commons Attribution Share Alike Norm (ODC-BY-SA),[44] which asks that changes and updates to the dataset are made public, that the original publisher is mentioned in the credits, that the source of the data is linked, that open formats are used, and that no further digital rights management (DRM) is applied. In the same way as the dcterms:license predicate is used to link the dataset to its license, so can the wv:norms predicate be used to link a dataset to a community norm specification.

7.10 SOFTWARE TOOLS

There are a number of commercially available and open source tools for storing and publishing linked data. Many tools also come with a query engine and RDFS inference. These include 3Store (a MySQL-based triple store from the University of Southampton), which has been extended by a spin-off company into 4Store (which is a database and query engine). Another option is the open source RDF database Mulgara,[45] which is written in Java with a SPARQL query engine.

There are a number of software frameworks (collections of tools and code libraries) dealing with various aspects of Semantic Web processing and Linked Data, including triple stores. For example, the Apache Jena Framework[46] includes a Java API for reading, processing, and writing RDF data in XML, N-triples, and Turtle formats; an ontology API for handling OWL and RDFS ontologies; a rule-based inference engine for reasoning with RDF and OWL data sources; an RDF triple store; and a SPARQL query engine. Sesame,[47] from Aduna Software, is another framework for processing RDF data, including RDF storage, inferencing, and SPARQL querying. It also offers transparent access to remote RDF repositories using the same API as for local access.

Commercial choices include the Oracle 11g database,[48] which uses its spatial data model to store the RDF graph; Allegro Graph,[49] which offers RDFS++ and Prolog reasoning, with a specific geospatial datatype; BigData,[50] which is an open source distributed database that also offers limited OWL inference; and OpenLink Virtuoso,[51] which is a hybrid relational and RDF data store, along with a linked data and Web application server.

However, the market is in flux, with new tools and features becoming available all the time, so we cannot cover all the possible choices here.

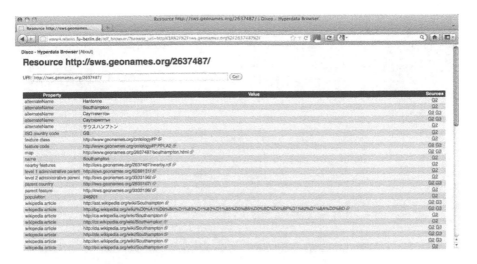

FIGURE 7.1 A screen shot of the Linked Data for http://sws.geonames.org/2637487 displayed in the Disco browser.

When discussing tool support, it is useful to be aware of how an end user may be accessing Linked Data. One common way is via a Linked Data-enabled Web browser. There are a number of these available, which can be divided into server-side applications that allow RDF data visualization and exploration in a standard Web browser, without the need to install anything on the client side; and tools that require a plug-in to be installed locally for a particular Web browser. One such browser that relies on a server-side application is the "hyperdata" browser Disco.[52] Disco sits on top of the Semantic Web Client Library,[53] which represents the whole Linked Data Web as a single RDF graph, and presents all the information about a particular resource in the Linked Data graph as an HTML page. It does this by displaying the values of the rdfs:label properties and dereferencing any property URIs. The user can click on the hyperlinks (which are triple predicates) to navigate between Linked Data resources. An example of the Linked Data for the GeoNames-coined URI for Southampton in the United Kingdom (http://sws.geonames.org/2637487) is shown in Figure 7.1.

Other examples of Linked Data browsers using server-side technology include

- Zitgist[54]
- Marbles[55] (a server-side application that formats RDF for HTML browsers using the Fresnel RDF Display Vocabulary[56] for displaying Semantic Web content in a human-readable way)

Other browsers that require installation of a software extension or plug-in include

- Tabulator[57] (an extension to the Firefox browser)
- OpenLink Data Explorer[58] (an extension to browsers such as Firefox, Safari, and Chrome that allows RDF/XML and RDFa browsing)

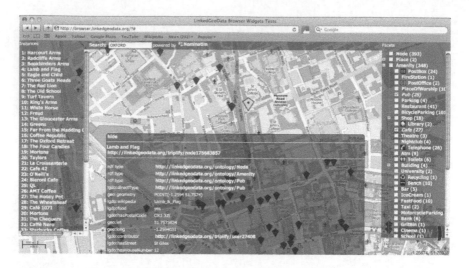

FIGURE 7.2   Linked Geo Data browser. (© OpenStreetMap, http://www.openstreetmap.org, contributors 2012, CC-BY-SA. http://www.openstreetmap.org/copyright)

A browser that displays RDF data against a slippy map, using OpenStreetMap data, is available from the LinkedGeoData[59] project. The Linked Data from OpenStreetMap is used to allow selection of certain geographical features ("facets"), which are then marked on the map, as in Figure 7.2.

There are also several Semantic Search Engines that crawl and index Linked Data to provide search facilities. These include the Semantic Web Search Engine (SWSE)[60] (Hogan et al., 2011), which indexes OWL, RDF, and RSS data; Sindice[61] (a lookup index for the Semantic Web); and Swoogle[62] (which searches through ontologies and instance data).

7.11   TESTING AND DEBUGGING LINKED DATA

Merea Maps has designed, created, and is ready to publish its Linked Data. Users will be able access it through a Linked Data browser or an application that sends SPARQL queries to Merea Maps' SPARQL endpoint. Merea Maps may also implement a Linked Data browser with mapping backdrop, similar to the GeoLinkedData browser. However, the final stage before publication must be to pass quality assurance—and indeed, testing should ideally be incorporated throughout, if not before, the Linked Data creation process. There are a number of levels on which the Linked Data must be checked for errors.

7.11.1   Syntactic Correctness

The W3C RDF Validation Service[63] can check RDF/XML validity, and the tool Eyeball[64] (which is part of the Jena framework) can identify specific syntax problems like properties and classes that have not been declared in a schema, prefixes with no namespace declared, ill-formed URIs, illegal datatypes, and untyped resources and literals, among others.

7.11.2   Correct URI Dereferencing

Regarding correct URI dereferencing, this is the question of whether the Linked Data server performs a 303 redirect on the URI correctly and eventually returns the description of the RDF resource. A number of tools can check this for you. The Vapour Linked Data Validator[65] provides a report about the HTTP requests and responses that took place as the URI was looked up. The URI Debugger[66] is a similar tool for debugging Linked Data sites and URI dereferencing. Hurl[67] makes basic HTTP requests, so you can see what is happening as your URI is dereferenced. RDF:Alerts[68] is another Web tool to validate RDF/XML; it dereferences the URI and checks the data is valid against its RDFS ontology. If all else fails, there is also the command-line-based tool cURL,[69] which transfers data in URL syntax to and from a server; this basic functionality allows you to check the fine detail of the HTTP communication to ensure the 303 redirects are operating correctly.

7.11.3   Content: Semantic Correctness

To test whether the Linked Data set is fully interconnected, it is worth using a Linked Data browser such as one of those discussed in Section 7.10. Check whether every URI in the dataset can be reached by browsing along the links. A fully interconnected Linked Data set will assist the Linked Data crawlers in indexing all of the dataset. Load the data into a tool that can perform semantic inferencing (a browser like Tabulator can do basic RDFS inferencing) and check that the data conforms to the domain and range restrictions, and no URI is inferred to be an instance of the wrong class. Hogan et al. (2010) reported on the types of errors frequently found in Linked Data, which included properties found in the object/value position of a triple; misuse of `owl:ObjectProperty` (see Chapter 9), when a datatype property should have been used instead; and the hijacking of properties defined elsewhere for the authors' own purposes. The worst example of this last transgression was when the standard property `rdf:type` was redefined.

7.11.4   Content: Semantic Completeness

For checking semantic completeness, convert the competency questions for your Linked Data into SPARQL (see Chapter 8) and query them using a SPARQL endpoint on your data to check that the data returns all the answers you expected. If it does not, you may be missing data resources, but more likely, you will be missing some links between resources. Check the individual triples in Turtle format as this is a more concise, easy-to-read syntax.

7.11.5   Expect the Worst

The analysis of errors in Linked Data from Hogan et al. (2010) suggests that Linked Data on the Web must be *expected* to be buggy, and just as browsers today are forgiving of errors in HTML and users do not necessarily believe everything they read on the Web, so will the Linked Data tools of the future have to be forgiving of malformed Linked Data. While tool developers will have to contend with Linked Data that does not conform to the Semantic Web best practices covered in this chapter, we must hope that the semantic value of the content however will be accurate. In the next chapter, we look at the issue of Semantic Spam and what happens when a data publisher has willfully produced incorrect data.

7.12   SUMMARY

This chapter has provided an introduction to how to publish Linked Data. We explained Berners-Lee's Linked Data Principles, in particular the importance of making URIs dereferenceable (i.e., it should be possible to look up the RDF description of the resource identified by the URI on the Web). We have taken the reader through the process of creating Linked Data: first, deciding what the Linked Data is about; second, examining the current GI data and identifying what should be included, left out, or modified; third, designing the RDFS ontology for the data, deciding on the purpose, scope, and competency questions for the dataset and reusing vocabularies if possible; fourth, minting the URIs for the data; and finally, generating data by running additional GIS queries if necessary. The RDF data can be published as a static RDF/XML file, as RDFa embedded in an HTML Web page, or using one of the many software tools to create a Linked Data view on a relational or RDF database. The publisher's own vocabularies should be made dereferenceable, and each URI should be provided with an `rdfs:label` to enable visualization tools to present a nice "human-readable" version of the resource. Metadata should be attached to the dataset, preferably using VoID, to refer to other access methods like RDF dumps and SPARQL endpoints, to specify what the data is about (e.g., by including the purpose, scope, and competency questions), and to include provenance and licensing information. Finally, to be really considered proper Linked Data, the publisher must link its data to other Linked Data on the Web for it to be discovered. This is the problem of data integration, and we address this in the next chapter.

NOTES

1. While in American English the # is known as a pound sign, the Linked Data community have adopted the British English word for it—the hash.
2. Other conventions are also used, and at the time of writing, none was universally accepted, so you need to be aware of the particular convention used by the publisher of the dataset you are interested in.
3. The *S* in RDFS stands for schema. The use of the word *schema* rather than ontology is due to RDFS's origin as a metadata language that pre-dates Linked Data. In this book, we have therefore emphasized that the product is an ontology by referring to it as an "RDFS ontology." More normally, you are as likely to hear people talking about "the RDFS" or "RDFS schema," meaning the ontology written using RDFS, whereas ontologies produced using OWL are commonly referred to as just "ontologies."
4. http://neologism.deri.ie/.
5. http://www.opencalais.com
6. http://www.ontos.com

7. http://www.cambridgesemantics.com/products/anzo-express
8. http://www.topquadrant.com/products/TB_Composer.html
9. http://xlwrap.sourceforge.net/.
10. http://www.w3.org/wiki/ConverterToRdf
11. A BLOB is a package of data with internal structure defined by the originator and normally requiring custom code to interpret.
12. http://virtuoso.openlinksw.com/whitepapers/relational%20rdf%20views%20mapping.html.
13. http://www4.wiwiss.fu-berlin.de/bizer/d2r-server/.
14. By "simplistic," we mean that a relational table represents an RDFS class, a table row represents an RDFS resource, and a column represents a predicate.
15. http://triplify.org
16. http://linkedgeodata.org
17. http://triplify.org/Documentation#h47-10.
18. http://geo.linkeddata.es
19. http://pegasus.chem.soton.ac.uk
20. http://www.kanzaki.com/works/2005/imgdsc/flickr2rdf
21. http://www4.wiwiss.fu-berlin.de/bizer/bookmashup/.
22. http://www4.wiwiss.fu-berlin.de/bizer/rdfapi/index.html
23. http://drupal.org
24. http://www.w3.org/2007/08/pyRdfa/.
25. http://www4.wiwiss.fu-berlin.de/pubby/.
26. http://code.google.com/p/djubby/.
27. http://vocab.deri.ie/void
28. http://www.opensearch.org
29. http://lab.linkeddata.deri.ie/ve2/.
30. http://www.hpi.uni-potsdam.de/fileadmin/hpi/FG_Naumann/projekte/btc2010/voidgen.jar
31. http://virtuoso.openlinksw.com/.
32. http://www.openprovenance.org/tutorial
33. http://openprovenance.org/model/opmo#ref-opm-v1.1
34. http://vocab.org/changeset/schema.html
35. http://trdf.sourceforge.net/provenance/ns.html
36. http://www.w3.org/2004/03/trix/swp-2/.
37. http://www.w3.org/2011/01/prov-wg-charter
38. http://www4.wiwiss.fu-berlin.de/bizer/ng4j
39. http://opendefinition.org/.
40. http://www.creativecommons.org
41. Note that this example uses the Creative Commons property cc:license rather than the dcterms one; either is permissible, although dcterms is more frequently found in Linked Data and hence may be the better option.
42. http://creativecommons.org/publicdomain/zero/1.0/.
43. http://vocab.org/waiver/terms/.
44. http://www.opendatacommons.org/norms/odc-by-sa/.
45. http://www.mulgara.org
46. http://incubator.apache.org/jena/index.html
47. http://www.openrdf.org
48. http://www.oracle.com/technetwork/database/options/semantic-tech/whatsnew/index.html
49. http://www.franz.com/agraph/allegrograph
50. http://www.systap.com/bigdata.htm
51. http://virtuoso.openlinksw.com
52. http://www4.wiwiss.fu-berlin.de/bizer/ng4j/disco/.
53. http://www4.wiwiss.fu-berlin.de/bizer/ng4j/semwebclient/.
54. http://dataviewer.zitgist.com/.

55. http://www5.wiwiss.fu-berlin.de/marbles/.
56. http://www.w3.org/2005/04/fresnel-info/.
57. http://dig.csail.mit.edu/2007/tab/.
58. http://ode.openlinksw.com/.
59. http://browser.linkedgeodata.org
60. http://swse.deri.org/.
61. http://sindice.com/search
62. http://swoogle.umbc.edu/.
63. http://www.w3.org/RDF/Validator/.
64. http://jena.sourceforge.net/Eyeball/.
65. http://validator.linkeddata.org/vapour
66. http://linkeddata.informatik.hu-berlin.de/uridbg/.
67. http://hurl.it/.
68. http://swse.deri.org/RDFAlerts/.
69. http://curl.haxx.se/.

8 Using Linked Data

8.1 INTRODUCTION

In Chapter 7, we covered the process of creating and publishing Linked Data from Merea Maps' own Geographical Information (GI). Now we move on to cover a number of issues surrounding the use of Linked Data: how to query, interlink, and create business value from it. We start by exploring the reasons why an organization with GI should consider investment in Linked Data publishing and discuss some Linked Data business models. Section 8.3 explains the RDF query language SPARQL, which can be used to construct queries into a triple store or published dataset via a SPARQL endpoint. The bulk of the chapter discusses the process of selecting links from Merea Maps' own dataset to third parties, including quality versus volume trade-offs between manually curated and automatically generated links and specifying the context in which a link might be valid. When we look at data integration through linking, it becomes apparent that the RDFS (Resource Description Framework Schema) language is limited in its ability to fully express the knowledge required to really know whether and when two things are the same or what other relationship exists between them. Motivated by these shortcomings, Chapter 9 then moves on to introduce methods of encoding more nuanced information in ontologies using the OWL Web Ontology Language.

8.2 BUSINESS MODELS FOR LINKED DATA

So, why should a GI data publisher go to the expense and effort of expressing its GI as Linked Data? While the Linked Open Data movement encourages the publication of data with an open license, for publishers who have spent years surveying and collecting data and whose business relies on selling data products, this is not always a realistic option. There are a number of ways to generate value from Linked Data, as explained by Scott Brinker[1] and Leigh Dodds.[2]

8.2.1 SUBSIDY MODELS

The subsidy model is the business model followed by, among others, the U.K. government's Linked Data initiative, the BBC, and the U.S. Census bureau. In this scenario, the publisher, usually a public sector, educational, or charitable organization, is funded (sometimes by donated time, such as in the case of GeoNames) to produce the Linked Data for public benefit.

8.2.2 INTERNAL SAVINGS

While not strictly a way of generating income, the process of structuring data in RDFS and linking to other external datasets can offer efficiencies in terms of improving the precision of data specification at the data collection stage, getting a clearer picture of the enterprise's knowledge, and opening up new opportunities for producing new data products more cheaply and quickly. In short, it helps the organization know what it knows.

8.2.3 TRAFFIC

Driving traffic to a Web site, and thus increasing the exposure of potential customers to your content and services, is a well-known model for revenue generation on the Web. An ever-growing number of organizations have developed Web Application Programming Interfaces (APIs) to encourage traffic to their site, and a similar logic can be applied to Linked Data. Publishing Linked Open Data allows Web crawlers to more easily find and index your data, which boosts your site's ranking in the major search engines such as Google. Brinker calls this "data-enhanced search engine optimization."

8.2.4 ADVERTISING

Advertising is a common business model for revenue generation on the Web, but how well does it apply on the Linked Data Web? At the time of writing, no publishers had yet taken the advertising route, so its success, or otherwise, remains to be seen. While we can clearly expect advertisements to appear on Web pages offering SPARQL endpoints and so on, it does not seem likely that embedding advertising into the raw Linked Data will take off. The data is presented to the end user only after processing by the application, which could easily identify and remove the advert. Also, targeted advertising, almost a requirement of today's ad campaigns, is difficult in Linked Data.

8.2.5 CERTIFICATION

Another way of making money from Linked Data could be for trusted authorities to sell certification services so that customers know that the data is of high quality, safe, and authentic. This has not yet been tried on the Linked Data Web, as it can only really be viable once other business models are in place and there is enough Linked Data to make authentication a necessity.

8.2.6 AFFILIATION

In this model, affiliate links are embedded in the data and displayed in end-user applications so that the retailer pays a commission to the Linked Data provider for every link that the user follows to its retail site or data. According to Brinker,[1] $6 billion of commissions are generated by standard Web affiliate marketing programs, so it seems likely that if this model can be applied to the Linked Data

Web, there is plenty of scope for revenue generation. Again, this is a business model that is likely to appear later in the development of the Linked Data market.

8.2.7 SERVICE BUNDLES AND AGGREGATION

Linked Data could be provided as part of a bundle of services to add value to other offerings or be offered as a sweetener to other data sales. By providing the links to third parties' data along with your own Linked Data, the purchaser of the Linked Data gains access to not only your data but also the third party's (in terms of integration, if not license). In effect, you are selling the fact that you have done all the data integration work for your customer; taking this idea further, you become a data aggregator.

8.2.8 BRANDING OR "LOSS LEADER"

Companies such as the *New York Times*[3] have published Linked Open Data, motivated initially by wanting to understand the technologies and position themselves as forward-thinking brands. The loss leader idea overlaps with branding as the Linked Data set can be provided for free as a way of promoting the brand and as an enticement for customers to buy the richer or higher-quality full dataset. This then moves us into the "free" part of the "freemium" business model.

8.2.9 SUBSCRIPTION ACCESS: CLIMBING THE PAY WALL

Subscription is a well-known business model, both for content and services; however, for Linked Data, if it is completely hidden behind a pay wall, this can reduce the likelihood of external data linking to it. Encouraging incoming links is hugely important because they increase the chances of discovery, provide a measure of popularity and trust of the dataset, and push it further up search rankings. At a minimum, users or software agents need access to metadata: a description of the data to know whether it is what they are seeking. The freemium model provides some data for free and charges for certain enhancements, such as more up-to-date data (or conversely, archival data) or more detailed, higher-quality, or a wider range of data. Certain access mechanisms, such as download, might be priced higher in exchange for the convenience. Charges can be subscription based, for a limited time period, or "pay as you go." The last case is interesting as there are of course many different ways to chunk up data: Could one even pay per triple?

Of all the business model options, we would argue that the freemium model is of the most interest to a GI publisher since it can provide the openness required to improve discoverability while also protecting the value of the data.

8.2.10 IMPLEMENTATION OF LINKED DATA BUSINESS MODELS

Cobden et al. (2011) noted that every Linked Data set currently published uses a loss leader or subsidy business model, and there are a number of technical hurdles to be overcome before the more attractive freemium models incorporating some form of paid access can be realized.

When a request is received by the server publishing the restricted-access Linked Data set, negotiations must take place between client and server to allow subscribers to be authenticated. The use of authentication protocols such as OAuth[4] or the World Wide Web Consortium (W3C) WebID protocol (Story and Corlosquet, 2011) have been proposed, although at the time of writing not yet tried in any commercial system. The idea is that if the user is successfully authenticated, the paid-for data can be served; if not, the server redirects nonpaying customers to the free data. New vocabulary is needed so that the free content can indicate that there is further content of interest behind the pay wall. There are also calls for standardization of how payment can be made, beyond the somewhat limited HTTP (Hypertext Transfer Protocol) "402 Payment Required" response code. We discussed how to describe data provenance in Chapter 7; however, there are not yet any well-established ways of automatically understanding license conditions and ensuring that they are met before providing the data.

In summary, it is very early days in the Linked Data market, and neither the technical nor the business details have yet been ironed out or tested in the wild.

8.3   SPARQL

We mentioned the SPARQL RDF query language several times in Chapter 7. Since SPARQL is taught in great detail by many other publications (Feigenbaum and Prud'hommeaux, 2008; du Charme, 2011), we do not go into great detail about the inner workings of the query-answering logic, but just provide the basics to enable readers to recognize and construct their own SPARQL queries and understand the results.

SPARQL, pronounced "sparkle," the recursively named SPARQL Protocol and RDF Query Language (Prud'hommeaux and Seaborne, 2008), is a W3C standard for querying RDF data. The standard also includes a protocol that defines how to query remote RDF data that has been published as Linked Data elsewhere on the Web. There are four main forms of SPARQL query: SELECT, ASK, DESCRIBE, and CONSTRUCT; these allow read operations (that is, to return results from the RDF graph).[5]

Most forms of SPARQL query start with one of these four keywords, which are applied to a set of triple patterns called a basic graph pattern. Like RDF, a SPARQL triple pattern consists of a subject, predicate, and object, but in SPARQL, one or more of these can be a variable. For example, in the following triple pattern, the object is the variable number, denoted with a ? (although $ is also valid to denote a variable).

```
@prefix mm_address: <http://id.mereamaps.gov.me/addresses>.
mm_address:0001 mm_address:hasHouseNumber ?number.
```

Note that we are using Turtle syntax here; the statement terminates in a full stop, and it is quite acceptable to have more than one or even all three of the subject, predicate, and object as variables. The triple pattern is matched against a subgraph of the RDF data, where the subject, predicate, or object in the RDF is substituted for the variable in the SPARQL triple pattern. In more complex queries, there will be a collection of triple patterns (which is called the graph pattern) that is matched against the RDF graph, for example:

```
{
?house mm_address:hasHouseNumber ?number.
?house mm_address:hasStreet "Troglodyte Street".
}
```

This graph pattern is the set of triples concerning houses (or actually any resource) that have a house number *and* are in Troglodyte Street. That is, if a variable appears in multiple triples within a graph pattern, it must be bound to the same value in all the triple patterns.

8.3.1   SPARQL SELECT

Now, we move on to some example SPARQL queries. First, to select a number of triples from the RDF graph, we use the following query structure:

```
1. base <http://data.mereamaps.gov.me/administrativeRegions.rdf>
2. prefix admin: <http://data.mereamaps.gov.me/administrativeRegions/>
3. select *
4. from <http://data.mereamaps.gov.me/administrativeRegions.rdf>
5. where
6. {
7. ?region    a admin:Region;
8.            :contains ?y.
9. ?y         a admin:City;
10.           admin:hasPlaceName 'Medina'.
11. }
```

Line 1 specifies the base Uniform Resource Identifier (URI) to which all URIs further in the query will be relative, and line 2 provides a shortcut to represent the namespace of Merea Maps' administrative regions using just the prefix admin. Line 3 is known as the "result clause" and specifies what we want to return; in this case, we want to select everything (denoted by *) from the RDF graph stated in line 4, according to the conditions in lines 7–10. Lines 7 and 8 are asking for instances of the class admin:Region that contain something (stored in variable ?y). Note that the SPARQL keyword a is an abbreviation for rdf:type, stating that we are looking for answers that are of the type admin:Region (line 7). Then, lines 9 and 10 further specify that the ?y we are looking for is an instance of the class admin:City, which has the name "Medina." So, the whole query will return all the information in the Administrative Regions RDF graph about the region that contains the city of Medina.

As with other data query languages such as SQL, the keyword DISTINCT can be added after the SELECT to only return unique results, avoiding duplicates. Solution modifiers can be added after the conditional where clause. For example, ORDER BY sorts results in ascending or descending order of one of the variables, LIMIT limits the number of results returned, and OFFSET n skips the first *n* results. The FILTER keyword allows you to exclude results whose values do not meet the specified constraints (e.g., to drop small areas from a query where the interest was in settlements with populations greater than 100), and the UNION keyword returns results that match exactly one of the mutually exclusive graph patterns presented in the where clause. For example, the following will return regions containing the city of Medina or the village of Ash Fleet:

```
1. base <http://data.mereamaps.gov.me/administrativeRegions.rdf>
2. prefix admin: <http://data.mereamaps.gov.me/administrativeRegions/>
3. select *
4. from <http://data.mereamaps.gov.me/administrativeRegions.rdf>
5. where
6. {
7. ?region   :a admin:Region;
8.            :contains ?y.
9.    {
10.          ?y  a admin:City;
11.              admin:hasPlaceName 'Medina'.
12.   }
13.   union
14.   {
15.          ?y  a admin:Village;
16.              admin:hasPlaceName 'Ash Fleet'.
17.   }
18.}
```

Finally, we should mention the OPTIONAL keyword, which allows results to be returned even if the part of the pattern within the optional clause is not matched. For example, the following query will return regions that contain any city and will return their names if the information is available:

```
1. base <http://data.mereamaps.gov.me/administrativeRegions.rdf>
2. prefix admin: <http://data.mereamaps.gov.me/administrativeRegions/>
3. select *
4. from <http://data.mereamaps.gov.me/administrativeRegions.rdf>
5. where
6. {
7.  ?region   a admin:Region;
8.            :contains ?y.
9.  ?y        a admin:City.
10. optional {?y ?hasPlaceName ?name.}
11.}
```

8.3.2   QUERYING MULTIPLE GRAPHS

The from clause used in the queries of Section 8.3.1 specifies what is known as the *background* graph. We can, however, specify multiple RDF graphs to query, using *named* graphs. We do this with the statement:

```
from named <uri>
```

and then use the GRAPH keyword within the query itself to indicate the graph to which we are referring. Querying multiple graphs at the same time allows us to aggregate data. For example, the following query will return administrative districts:

```
1.  prefix admin: <http://data.mereamaps.gov.me/administrativeRegions/>
2.  prefix geo: <http://www.w3.org/2003/01/geo/wgs84_pos#>
3.
4.  select distinct ?graph_id ?property ?hasValue
5.  from <http://data.mereamaps.gov.me/administrativeRegions.rdf>
6.  from named <http://data.mereamaps.gov.me/administrativeRegions.rdf>
7.  from named <http://postcodes.mm.gov.me>
8.  where
9.  {
10.   graph <http://data.mereamaps.gov.me/administrativeRegions.rdf>
11.   {
12.     ?city  a admin:District;
13.            geo:lat ?lat;
14.            geo:long ?long.
15.   }.
16.   graph <http://postcodes.mm.gov.me>
17.   {
18.     ?city  geo:lat ?lat;
19.            geo:long ?long.
20.   }.
21.   graph ?graph_id
22.   {
23.     ?city  geo:lat ?lat;
24.            geo:long ?long;
25.            ?property ?hasValue.
26. }
27. }
```

This query asks for unique solutions of the graph URI, any property and any value of that property to be returned (line 4) from the named graphs of Merea Maps' administrative regions RDF graph (line 6) and the Merean Mail postcodes RDF graph (line 7), where the solution meets the criteria that in the administrative regions graph, the resource is a city with a certain latitude and longitude (lines 11–15), and in the postcodes graph, it has the same latitude and longitude (lines 16–20). Finally, lines 21–26 allow the return of all the information (any property and its value) from each graph about those districts present in both the administrative regions and post-codes dataset.

8.3.3   SPARQL ASK, CONSTRUCT, AND DESCRIBE

The SPARQL keyword ASK returns true or false depending on whether there are any matches to the query. For example, the following asks if Medina is in the administrative region of North Merea:

```
1. prefix admin: <http://data.mereamaps.gov.me/administrativeRegions/>
2. ask
3. {
4. ?medina  admin:hasPlaceName 'Medina';
5.          admin:isContainedIn ?region.
6. ?region  a admin:administrativeRegion;
7.          admin:hasPlaceName 'North Merea'.
8. }
```

Note that this particular query does not tell us whether Medina also happens to be in other administrative regions, and we are assuming that the place name provides a sufficiently unique identifier. The ASK keyword is particularly useful for deciding if the two datasets have any information in common.

The CONSTRUCT keyword returns a new RDF graph, so if we wanted to create a graph of those cities (that are administrative units) in Merea Maps' dataset, using Merean Mail's vocabulary (e.g., using the term "hasPostCity" rather than "hasPlaceName"), we could make the following query:

```
1. prefix admin: <http://data.mereamaps.gov.me/administrativeRegions/>
2. prefix postcodes: <http://data.mm.gov.me/postcodes.>
3. construct {?city postcodes:hasPostCity ?name}
4. where
5. {
6. ?city  a admin:City;
7.        admin:hasPlaceName ?name.
8. }
```

This is effectively a way of converting data from one schema to another.

Finally, the DESCRIBE keyword instructs the query to return any information about a data resource that the data publisher would like to return. Often, this is specified in the concise bounded description included in the Semantic Sitemap or Vocabulary of Interlinked Datasets (VoID) metadata (see Section 7.6); however, it could be the named graph or minimum self-contained graph. Since there is no standard definition of what will be returned, describe queries are not interoperable. For example, the following query will return the "description" (that is, all the information the publisher thinks you need to know) about the administrative region of North Merea:

```
1. prefix admin: <http://data.mereamaps.gov.me/administrativeRegions/>
2. describe ?northmerea
3. where
4. {
5. ?northmerea a admin:administrativeRegion;
6.                admin:hasPlaceName 'North Merea'.
7. }
```

8.3.4   GEOSPARQL

We touched on the query language GeoSPARQL in Section 6.4 and its representation of geometry, and in Section 6.6 also discussed the Region Connection Calculus 8

(RCC8) spatial relations that it uses. Here, we show how these are used in the query language itself, which defines spatial extensions to SPARQL and is about to be published as an Open Geospatial Consortium (OGC) standard in its own right (Perry and Herring, 2011). GeoSPARQL consists of

- An RDF/OWL vocabulary for representing spatial information
- A set of functions for spatial calculations
- A set of query transformation rules

GeoSPARQL models some spatial concepts in an OWL ontology; for example, it has SpatialObject as its top-level class, with a direct subclass of Feature. There are also classes for geometry objects and RDFS datatypes for representing geometry data (such as points, lines, and polygons as discussed in Chapter 6). The property `geo:hasGeometry` links a feature with a geometry that represents its spatial extent and `geo:asGML` relates a geometry to its Geography Markup Language (GML) serialization. A feature can have multiple geometries. This step, of separating the concept of the object from the concept of its spatial representation (footprint, or point location), is very important as it allows us to separate the logical reasoning we want to carry out on the semantics from the geometric calculations we need to carry out on the spatial data. There are also a number of properties for spatial relationships, with three different families of spatial relations included as part of the standard: the Simple Feature Relations model, the RCC8 spatial relations, and the Egenhofer 9 Intersection model relationships. These can be used directly in SPARQL triple patterns to test whether spatial relationships exist between two instances of `geo:SpatialObject`.

A number of functions are available in GeoSPARQL for spatial calculations, for example, `geof:distance`, which returns the shortest distance in units between any two points in two geometric objects as calculated in a particular spatial reference system. Also, `geof:buffer` returns a geometric object that represents all points whose distance from the geometric object is less than or equal to the given radius. Other functions that can be used are `geof:convexHull`, `geof:intersection`, `geof:union`, `geof:difference`, `geof:envelope`, and `geof:boundary`.

Finally, GeoSPARQL includes a set of query transformation rules (specified in the Rule Interchange Format), which expands a triple pattern using a spatial predicate into a set of triple patterns plus a Boolean query function. That is, the rules map each of the spatial relations onto a function, which actually does the calculations on the geometries to see whether the spatial relationship holds.

Using GeoSPARQL, we can, for example, pose the query "Find the pubs in Merea" (where the "in" means "spatially contained in"):

```
1.  prefix geosparql: <http://www.opengis.net/def/geosparql/>
2.  prefix mm: <http://mereamaps.gov.me/topo/>
3.  prefix admin: <http://mereamaps.gov.me/administrativeRegions/>
4.  select distinct ?pub where {
5.    graph <http://data.mereamaps.gov.me> {
6.    <http://data.mereamaps.gov.me/0001> a admin:Country;
```

```
7.         admin:hasPlaceName 'Merea';
8.         geosparql:hasGeometry ?merea_geom.
9.  ?pub a mm:Pub ;
10.        geosparql:hasGeometry ?pub_geom.
11. ?pub_geom geosparql:sfWithin ?merea_geom.
12. }
13.}
```

Line 4 selects unique answers to the query, as returned in the ?pub variable. For simplicity, we assume that all Merea Maps' data (both administrative regions and topographic) can be accessed from one big graph as stated in line 5. The instance 0001 (which is a Country with place name "Merea," lines 6 and 7) has a geometry stored in the variable ?merea _ geom (line 8). Lines 9 and 10 find instances of the class Pub with a geometry to be stored in the variable ?pub _ geom. Finally, in line 11, the GeoSPARQL sfWithin function is applied between the values in the ?pub _ geom and ?merea _ geom variables.

Instead of having to find the geometries of each spatial object and using a function to compare them, the query rewriting of GeoSPARQL allows you to query the two spatial objects directly using a spatial relation. So, you can write this simpler query, and behind the scenes, it will be rewritten into the previous query:

```
1. prefix geosparql: <http://www.opengis.net/def/geosparql/>
2. prefix mm: <http://mereamaps.gov.me/topo/>
3. prefix admin: <http://mereamaps.gov.me/administrativeRegions/>
4. select distinct ?pub where {
5.    ?pub geosparql:sf-within <http://data.mereamaps.gov.me/0001>.
6.    }
7.}
```

8.3.5 USING SPARQL TO VALIDATE DATA

The other important use of SPARQL is to validate data. When preparing RDF datasets, Merea Maps wants to make certain that they have not missed any data. For example, if a new set of houses is built in Ash Fleet, data needs to be gathered so that they are all assigned a house number, street name, and postcodes. Since RDFS is based on the open world assumption "just because you haven't said it, doesn't mean it isn't true," if one house is missing its house number, the data will still be valid under RDFS. The reasoner simply concludes that the house has a number; we just do not know what it is. To catch such problems in the data, we can use SPARQL to check that every house does indeed have a known house number.

The general model for validating data is to specify some ontology for expressing rules and constraints on the data and use it in combination with SPARQL to query the data to see if the constraints hold. Currently, there are two competing ways to validate RDF data: SPIN[6] and the Pellet Integrity Constraints Validator[7] (ICV). SPIN is an RDF vocabulary for specifying business rules and constraints, which has been published as a W3C member submission (Knublauch, 2011) and is supported by the Topbraid Composer editing and query tool. Pellet ICV uses an OWL integrity constraint ontology and translates this automatically into SPARQL queries to validate RDF data. If the query results indicate integrity constraint violations, Pellet ICV can also provide automatic explanations of why this has happened in order to assist debugging and data improvement.

8.4 LINKING TO EXTERNAL DATASETS: TYPES OF LINK

In Section 8.5, we outline the process for designing and creating the data links, but first we look at the big picture of the types of links that can be created and when they are appropriate to use. There are four different cases under which RDF data can be linked; three occur when there is correspondence at the instance level, between resources in the two different datasets, and the other is at the class level, between the vocabularies describing the two datasets.

8.4.1 CORRESPONDENCE BETWEEN CLASSES

We can link classes or properties in the two RDFS ontologies, which Heath and Bizer (2011) call *vocabulary links*. If we believe the classes are equivalent—so every instance in one class will also be an instance in the other (although the two parties may know about different instances)—we can use the `owl:equivalentClass` relationship. Note that this does not make any statements about instances within the classes matching at all. Similarly, `owl:equivalentProperty` can be used to state that a property in one vocabulary is equivalent to the property in another. If we cannot be sure that the classes or properties match exactly, `rdfs:subClassOf`, `rdfs:subPropertyOf`, `skos:broadMatch`, or `skos:narrowMatch` can be used to express a looser relationship.

8.4.2 CORRESPONDENCE BETWEEN INSTANCES: IDENTITY

Another type of link is the *identity link*, which specifies an equivalence relationship between two data resources, using the `owl:sameAs` predicate. It is straightforward to find this kind of link where the data is overlapping or the same but can be difficult to spot when the data is completely different but describes the same thing. For example, a Building may be described by its spatial footprint by one person, but by its address by another person. A third person, who wants to know the location of the address, will want to match these two types of data and would find the `owl:sameAs` link between the two URIs very useful. We talk more about how to find this kind of match in Section 8.6.

If Merea Maps is not sure whether anyone else has already minted a URI to identify the thing in which it is interested, it is quite acceptable for it to create its own URI and, when ready, to seek out other URIs referring to that entity. These URIs are known as *URI aliases*. Since Merea Maps' URI contains its namespace, the information that it has published about that entity can easily be traced back to Merea Maps through dereferencing, which is useful for others on the Web who want to know what Merea Maps is saying about that thing. Multiple URIs about the same thing, linked through the `owl:sameAs` predicate, also convey a very strong advantage on the Linked Data Web: robustness. As with the traditional Web, there is no central point of failure, and the decentralized approach means there is no administrative burden of centralized organization required to assign unique URIs to each individual entity (as well as the impossibility of defining to everyone's satisfaction exactly what the entity is).

8.4.3   CORRESPONDENCE BETWEEN INSTANCES: DIFFERENCE

If the information provided is the same but the things the two URIs are describing are different, we need to be more careful than merely assigning an `owl:sameAs` link. For example, GeoNames' "Southampton" (in the United Kingdom) and Ordnance Survey's Administrative Geography's "Southampton" both have the same name, but GeoNames refers to the City, a settlement, and Ordnance Survey refers to the Unitary Authority Area. For certain purposes, these two could be considered the same (if we were just interested in the general location of Southampton, say), but for the specific context of local government, we might need to link more accurately. While the GeoNames' Southampton (Settlement) does lie within the County of Hampshire, Ordnance Survey's Unitary Authority Area is not an administrative part of the County of Hampshire. Therefore, we need to specify the context in which the two entities can be considered the same; see Section 8.8 for more details of how to do this.

8.4.4   CORRESPONDENCE BETWEEN INSTANCES: OTHER RELATIONSHIPS

It is more usual, however, that there is some relationship other than similarity or identity between two resources from different Linked Data sets. For example, the Isis Tavern from Merea Maps' data *sells* Midnight Lightning Beer that comes from the Best Beer dataset. Heath and Bizer (2011) refer to this as a *relationship link*. As with the previous case, some relationships may only be valid in certain contexts of use.

8.4.5   ENCODING OUTGOING LINKS

An outgoing, or *external*, link can be encoded using a triple where the subject of the triple is in Merea Maps' dataset, and the object of the triple is located in another dataset under a different namespace. The predicate may reside in Merea Maps' namespace or in the third party's or alternatively come from yet another vocabulary. The following example shows internal links from the URI for the Isis Tavern pub in Merea Maps' data to other Merea Maps' data (line 7, in this case a literal) and external links to Merean Mail's address data (line 8) and to GeoNames data (line 9):

```
1   @prefix rdf: <http://www.w3.org/1999/02/22-rdf-syntax-ns#>.
2   @prefix foaf: <http://xmlns.com/foaf/0.1/>.
3   @prefix mm: <http://mereamaps.gov.me/placesOfInterest/>.
4   @prefix postcodes: <http://mereamail.gov.me/addresses/>.
5
```

```
6   <http://data.mereamaps.gov.me/placesOfInterest/0012>
      rdf:type mm:Pub ;
7     mm:hasPlaceName "The Isis Tavern" ;
8     postcodes:hasAddress <http://data.mereamail.gov.me/12345>
9     foaf:based_near <http://sws.geonames.org/7290621/>.
```

When the object of the triple is dereferenced, the third-party server provides all the information about that remote resource. That resource in the other dataset will be described by many other triples, which may in turn reference resources in even farther-flung datasets. This is the way in which the whole Web of data is linked, and by following these links (known as the "follow-your-nose" approach), a user can browse or a spider can crawl through the Web of Linked Data.

8.4.6 ENCODING INCOMING LINKS

If Merea Maps has identified links outgoing from its dataset, the inverse links may well also be valid, and Merea Maps can include these in their dataset, even though the subjects of the triples belong to the third-party dataset. This is particularly useful if the third-party dataset is one of the hubs in the Linked Data Cloud of Figure 2.1, as it is unlikely that the owners of the popular hub data will be able to create links out to all possible datasets that are relevant to them. Building on the previous example from Section 8.4.5, as well as encoding the outgoing links saying that the Isis Tavern has address postcodes:12345 and is based near geonames:7290621, we can also add in that postcodes:12345 isAddressOf The Isis Tavern (line 6) and geonames:7290621 is based _ near The Isis Tavern (line 7).

```
1   @prefix rdf: <http://www.w3.org/1999/02/22-rdf-syntax-ns#>.
2   @prefix foaf: <http://xmlns.com/foaf/0.1/>.
3   @prefix mm: <http://mereamaps.gov.me/placesOfInterest/>.
4   @prefix postcodes: <http://mereamail.gov.me/addresses/>.
5
6   <http://data.mereamail.gov.me/12345> postcodes:isAddressOf
<http://data.mereamaps.gov.me/placesOfInterest/0012>.
7   <http://sws.geonames.org/7290621/> foaf:based_near
<http://data.mereamaps.gov.me/placesOfInterest/0012>.
```

Note that although we have used foaf:based _ near to demonstrate how external links can work, this example is a little contrived; it is not usual to state that the town geonames:7290621 is near the Isis Tavern pub. One would normally only consider the pub to be near the town unless the pub was exceptionally famous.

8.5 LINK DESIGN PROCESS

As we have already mentioned, since search algorithms operate on the principle of assigning popularity to a Web page (or Linked Data node) based on the number of incoming links weighted by the linking pages' or nodes' own popularity, it is very important to link one's datasets to the rest of the Linked Data Web to increase discoverability.

8.5.1   STEP 1: SPECIFY THE PURPOSE, SCOPE, AND COMPETENCY QUESTIONS

Linking datasets can be treated in the same way as authoring a new dataset, following the steps set out in Chapter 7: First, the purpose and scope of the link ontology must be specified, and a set of competency questions can be developed (which can be expressed in SPARQL) for testing the links. Again, using Merea Maps as our example, Merea Maps' link architect wants to link the administrative regions' Linked Data set that they have just published to other datasets. Merea Maps could have two quite distinct purposes for this linking: One is to improve *discovery*, in which case identity and similarity links should be created, while the other would be for *repurposing* if Merea Maps had found some other data sources that it wanted to pull into its own dataset, thus enlarging the set of data available to its customers in a specific application, in which case relationship and vocabulary links would be required. So, the Discovery Purpose could be stated as: "To describe correspondences between Merea Maps' administrative regions and Merean Mail addresses to enable users to identify the administrative regions in which addresses lie." In contrast, the purpose of links created for a Repurposing use case could be stated as: "To describe Merean Mail addresses in terms of Merea Maps' vocabulary." This would be needed if, for example, we wanted to use Merean Mail data within an application that only handled Merea Maps' administrative data. It should be noted that in many cases both discovery and repurposing will be desirable, but it is still useful to recognize the differences and explicitly state these purposes.

The purpose is, in effect, stating the *context* in which the links are valid, and the scope provides the *coverage*. The scope of our first Discovery use case could be stated as, "to cover the description of addresses within Merean administrative regions." Classes that fall outside the intersection between the two datasets, such as Post Office Box Numbers, need not be modeled in the linking ontology. In the Repurposing use case, the scope of the Linked Data sets would again be the intersection of the Merea Maps and Merean Mail datasets but limited only to terminology used by Merea Maps.

8.5.2   STEP 2: IDENTIFY DATA SOURCES

As with the creation of a stand-alone ontology, the second step in Link Ontology creation is to examine the two datasets and two ontologies describing those datasets, also taking into consideration other vocabularies that may provide useful relationships for reuse. Let us take the Discovery use case mentioned: Merea Maps wants to link to Merean Mail data to make its own data more easily accessible on the Linked Data Web. In this case, Merea Maps needs to look at its administrative regions' Linked Data, which was extracted from their database of Table 7.1 and modeled as Linked Data in Chapter 7. Classes and example instances are shown in Table 8.1, including a more detailed explanation of the terms used, to help us when we need to align them with the Merean Mail addresses.

Merea Maps then compares its own Linked Data against the Merean Mail Linked Data, as shown in Table 8.2.

TABLE 8.1
Classes and Example Instances in Merea Maps Administrative Regions' Linked Data

Class Name	Example	Linking Considerations
Administrative Area	All of the following are administrative areas	An umbrella class that need not be directly linked.
Country	Merea	Since all of the Merean Mail addresses are within Merea, this class and single instance need not be directly linked.
Region	North Merea	A region is a subdivision of the country used for gathering statistics, and each has a Regional Statistics office covering a number of local authorities.
District	Greater Medchester	Districts are local administrative units.
County	Mereashire	Counties are alternative administrative units for areas that do not fall under a District.
City	Medina	Big cities like Medina have their own city council, so "City" is the administrative area for this. Even if Medina is spatially within Mereashire, it will not be administered by the Mereashire County council.
Parish	Ash Fleet Parish	Parishes are subdivisions of Counties or Districts and represent the most local level of government. This class refers to administrative, not ecclesiastical, parishes.

TABLE 8.2
Classes and Example Instances in Merean Mail Address Linked Data

Class Name	Example	Linking Considerations
Mail Box Number	171	Has no overlap with data in Merea Maps' administrative regions.
Organization Name	Isis Breweries Ltd	Again, will not be directly linked to.
Postcode	ME12 345	Not directly linked to.
Postcode Area	BLOB	May be used in geometrical calculations to see if two instances are spatially the same.
Postcode District	BLOB	This is a subarea of the postcode area that bears a close relation to vernacular place names and hence may be more useful than Postcode Area for geometrical calculations to see if two instances are spatially the same.
County	Mereashire	Relates to Merea Maps' "County."
Locality	Ash Fleet	Relates to Merea Maps' "Parish."
Dependent Locality	Medina	Relates to Merea Maps' District or City.
Road	Troglodyte Street	Not directly linked to.
Dependent Road	Big Street	Not directly linked to.
Building Number	39	Not directly linked to.
Building Name	The Isis Tavern	Not directly linked to.

Although we have identified some concepts from each Linked Data set that might be related to each other, as we can tell from the descriptions of the classes, we cannot immediately say that Merea Maps' Parish is the same as Merean Mail's Locality—they may not be spatially identical, and they do not have the same purpose. It is clear that many of the links will be relationship links on the instance level; so, the next step in the link design process is to decide what these predicates will be.

8.5.3   STEP 3: SPECIFY YOUR RDFS ONTOLOGY

The step of specifying the RDFS ontology is primarily a task of choosing relationships to use as link predicates. It is likely to be an iterative process with the Link Data generation step (see more about link discovery and creation in Section 8.6). For the first iteration, Merea Maps identifies the classes in the two Linked Data sets that are candidates for linking together (either at the class level or, more probably, at the instance level.) It then considers some options for potential link predicates, such as `owl:sameAs`, `owl:EquivalentClass`, or spatial relationships from the OGC GeoSPARQL standard, where `geo:rcc8-po` (partially overlapping) will be particularly useful, and it coins some new relationships, such as `isAdministeredBy`, `hasPostCodeArea`, and `hasAddressLocality`. The namespace of these new relationships should belong to Merea Maps; since Merea Maps is doing the link generation, it should retain control over the new predicates and their namespace. However, it is better if the new predicates are awarded their own namespace as they belong to a different ontology than that used by the stand-alone Linked Data sets. This helps to maintain the modularity of the Linked Data sets, so Merea Maps' administrative regions data can be reused without including these additional link predicates, if necessary.

Step 4 of the stand-alone Linked Data creation process, to mint new URIs, is not needed unless intermediate resources are required to bridge the gap between the two datasets. Step 5, to generate the data, is more complicated than the stand-alone process, as this is now a question of link discovery and creation, and hence the entire Section 8.6 is devoted to this issue.

8.6   LINK DISCOVERY AND CREATION

8.6.1   MANUAL LINK CREATION

Manual link creation is really only suitable for small datasets or high-quality datasets for which editorial oversight is paramount. It can also be useful for evaluating the success of automated methods. A Semantic Web search engine such as Sindice,[8] SWSE (Semantic Web Search Engine),[9] Falcons,[10] or SameAs.org[11] can be used to search for URIs of resources that are candidates for linking. Often, semiautomatic methods are used; for example, if there is a single or limited number of third-party datasets to link to, queries can be crafted to identify the most likely candidates within those datasets (for example, where the `rdfs:labels` are the same). Note that the link should go to the resource itself, not the document describing the resource (recall that resources typically have URIs containing the word "id").

8.6.2 Automatic Discovery and Creation

We use the terminology *matching* to mean seeking out equivalences, at either the class or the instance level, versus *linking*, for which we are trying to relate two classes/instances by some other relationship. Both of these cases are commonly referred to as linking, but we feel the distinction is worth making. In some ways, matching is the simpler case, so we address it first.

Currently, people use top-down matching rules, such as string-matching or hierarchical correspondence, to match two instances, along with various, usually heuristically derived, similarity measures. For example, when matching two places, we could combine the result of string matching their two place names, with some distance measure between their latitude and longitudes; of course, this is made a lot easier if the two datasets are using the same set of properties, so ontology matching is also required. An alternative to try is *bootstrapping*, a bottom-up approach that uses a small set of manually matched or linked instances to derive a more general matching/linking rule for similar cases. So, for example, if I have stated that mm:Mereashire owl:sameAs dbpedia:MereaCounty and so on for several other counties, I could derive (a) that mm:County owl:equivalentTo dbpedia:County, and (b) that it would be worth doing string matching on other counties in the two sets of counties.

There are several tools that can assist with link discovery, for example, Silk,[12] which is a graphical tool for identifying links between one RDF dataset and another on the Linked Data Web. Another tool that may be of use is the Linked Data Integration Framework,[13] which works with a Silk link-mapping specification and handles the disparities that can occur when some datasets are RDF/XML dumps only, while others are offered via SPARQL endpoints. The LIMES[14] (Link Discovery for Metric Spaces) tool has both a stand-alone option and a Web interface that works with SPARQL endpoints. LIMES works by finding a set of examples in the target dataset and matching each of the instances in that target dataset to their nearest example. Next, the distance between each target example and all the source instances is calculated, and any obvious mismatches (which have a large distance) are filtered out. Then, the actual distances between the source instances and the most likely target instances are calculated. This approach reduces the search space and number of similarity calculations that have to be carried out. Finally, the source and target instances with the highest similarity are output in N-Triples format. Another approach to link discovery is to use Bayesian belief networks, for example, the RiMOM[15] (Risk Minimization-Based Ontology Mapping) tool; however, this is limited to demonstration with a benchmark dataset only.

At the time of writing, automated link discovery was still a very immature area and the subject of ongoing research. Most of the tools described have significant limitations with accuracy, scale, or robustness and are for the most part still emerging from the universities where they were developed. They are therefore not yet mature enough to offer commercial-quality solutions to the problem of link creation. Nevertheless, they are indicative of how the technology is developing. For specific datasets, the advice is still to write one's own link discovery scripts based on knowledge of the datasets as this will produce higher accuracy than these more general tools.

8.7   ENCODING CONTEXT: AN ATHEIST'S VIEW OF WEB IDENTITY

Unlike many statements in the literature, we do not support a "God's eye view" on the world. That is, we do not believe that there is *any* overarching view of identity—there *is* no common view of Ash Fleet, say, to which everyone will agree. There are multiple contexts, and any agreement can only be made within a specific context. Some contexts may be quite general, others more specific, but there is no absolute agreement. Linked Open Data practitioners have struggled with this issue and usually cite the example of a person as the clearest case of the God's eye view of identity: Surely there is some representation of a person that everyone can sign up to? Obviously, not everyone needs to bother with all the attributes about a person; some will be interested in his or her e-mail address, while others want to know what papers they published, but there seems to be general agreement about the identity of a person. This absolute representation is paralleled in the idea of bona fide boundaries or objects in the geographic domain, but it is a fallacy. For example, we need context to answer questions like: Is Luke Skywalker a person? What uniquely identifies a person? (Mozart might be a person, but he does not have a Social Security number or e-mail address.) Since there is no context-free definition of an instance's identity, every matching decision is in fact made within a context.

In the mentioned example of Ash Fleet, instances can be matched or more linked generally only under a particular context. So, the next question is: If I am to link my data to other nodes on the Semantic Web, how do I specify the context in which those links are valid?

There are a number of different ways to encode context: both explicitly, through technical mechanisms of reification, named graphs, or what we term Named Triples; and implicitly, by characterizing the "landscape" of links, that is, the clustering of nodes in the graph, which may be tightly clustered or fall under different namespaces.

We mentioned the concept of reification for encoding metadata about RDF in Section 5.5.1, and this is one mechanism that people have turned to for describing the context in which the RDF graph is valid. Recall that reification specifies an `rdf:Statement`, which has a subject, predicate, and object, and allows other information, namely, metadata, to be added about that statement. However, as we have seen, the technique of reification has several drawbacks. First, it makes querying difficult as the triple graph model has been broken. Second, it buries metadata within the data itself, potentially making the volume of data balloon and forcing a user to deal with both data and metadata at the same time, which can make constructing queries very ugly. An alternative to reification is the technique of Named Graphs (Carroll et al., 2005), which names multiple RDF graphs within a single data repository or document with their own URIs, so that they can be individually referenced, allowing context information to be applied to each named graph individually.

Named graphs can be taken to the extreme of granularity, that is, to assign a context (in the form of a URI) to each resource individually, or even to each triple—a "Named Triple" if you will. Although Named Graphs are not part of the RDF standard, they are mentioned in the SPARQL 1.0 standard, which specifies that graphs can be explicitly named within a SPARQL query (see Section 8.3.2), and in

SPARQL 1.1, which explains how Named Graphs can be used in SPARQL Update[16] or the REST alternative, the Uniform HTTP Protocol for Managing RDF Graphs.[17]

Let us think about how Merea Maps could use Named Graphs to store their context, for example, the triple stating that the Isis Tavern is near the River Isis:

```
@prefix mm: <http://mereamaps.gov.me/topo/>.
mm:isis_tavern mm:near mm:river_isis.
```

Merea Maps decides to store this triple in a separate named graph so that it can explain its context (in this case, *what* is meant by "near," but the context could equally explain *when* the fact is valid). Merea Maps has used the "near" predicate to state that Medina is near Merea City elsewhere in its data, which is obviously applying "near" in a different context. Although there are several ways to encode named graphs, such as the TriG syntax for Named Graphs (Bizer and Cyganiak, 2007), which is an extension of Turtle, the only way that is self-contained in RDF is to use the SPARQL 1.1 service description.

In this way, as Merea Maps provides data via its SPARQL endpoint, it includes a description about the context in which its triple is valid (and whatever additional information Merea Maps chooses to add in here, such as copyright notices). Note that the triple has been loaded into the triple store under the specific named graph <http://mereamaps.gov.me/topo/graph/0012_near>:

```
@prefix sd: <http://www.w3.org/ns/sparql-service-description#>.
@prefix mm: <http://mereamaps.gov.me/topo/>.
mm:topo_endpoint
    a sd:Service;
    sd:url <http://mereamaps.gov.me/sparql>;
sd:availableGraphDescriptions [
    a sd:GraphCollection;
    sd:namedGraph mm:isis_near_graph;
];.
mm:isis_near_graph
    a sd:NamedGraph;
    sd:name <http://mereamaps.gov.me/topo/graph/0012_near>;
    mm:has_context 'Small scale';
    mm:copyright 'Merea Maps';.
```

Note that for many resources, however, it is not necessary to be so picky, and mostly, Merea Maps assigns a named graph on a per resource basis, so that the concise bounded description (i.e., all its literal properties, related blank nodes, and properties that link to related resources; see Section 7.6.1) of the resource is contained in one named graph.

All these techniques rely on being able to explicitly specify the context of use. However, context is often an emergent property, as shown in Figure 8.1, where URIs A, B, and C are tightly coupled (A links to B and vice versa), whereas D links to A, but A does not link back to D. This makes A, B, and C more of an emergent cluster than D, indicating that the links between A, B, and C are based on the same

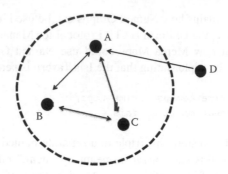

FIGURE 8.1 Links between URIs.

context, whereas D links to A for a different reason. (Note that these links need not, and probably will not, be `owl:sameAs`, but other, more complex links like "controls," "is located in," etc.). Therefore, although it is clearly better to specify context explicitly, context can also emerge implicitly from links already present on the Semantic Web.

At the macro scale, there are also links between entire datasets—the only thing the modularization of data into a prepackaged dataset is really giving us is the knowledge that all of the nodes within that dataset have the same context. It is worth referring back to the vision of a properly joined Semantic Web, where there should be little technical difference between an interdataset link and an intradataset link; the only differences should be semantic.

8.8 LINK MAINTENANCE

Dealing with the situation of identifying and managing broken links is another relatively unexplored area of Linked Data. Links tend to break when one or another (and sometimes both) of the datasets change. Therefore, the most obvious approach to avoiding the problem of broken links is via a system of notification when a dataset changes. According to Leigh Dodds,[18] there are four categories of information that could be notified: dataset notifications, when a new dataset has been added or updated; resource notifications, which detail which resources have been added or changed within the dataset; triple notifications, which provide information about the individual triples that have changed; or graph notifications, when modifications have been made to certain named graphs within the dataset. Notifications can be push or pull. Pull mechanisms include subscribing to feeds, using Linked Data crawlers, or querying datasets repeatedly to identify changes. Push mechanisms work by the data consumer subscribing to a system to which the data provider publishes information about its changes.

A number of ontologies have been published to describe the frequency of data updates, for example, the Dataset Dynamics Vocabulary[19] or time-ordered sets of update events such as the DSNotify Eventset Vocabulary (Popitsch and Haslhofer, 2010). While DSNotify takes a resource-centric view of changes, the Talis Changeset

Vocabulary[20] takes a triple-centric perspective. It is a set of terms for describing changes to triples, where the "Changeset" is the difference between the old dataset and the updated one. There are also ontology-level change description mechanisms, such as the OWL 2 Change Ontology (Palma et al., 2009), which allows the ontology publisher to describe how one version of an ontology differs from its predecessor. The Protégé ontology authoring tool offers the Change and Annotation Ontology (Noy et al., 2006), which allows the data provider to specify changes to the ontology, as well as version control information such as author of the change, timestamp, and other annotations.

The DSNotify framework implemented by Popitsch and Haslhofer (2010) monitors Linked Data sources and notifies applications consuming that data when the dataset has changed. Links may be either structurally broken (when the object resource is no longer retrievable) or semantically broken (where the link is semantically incorrect, for example, two resources are linked with owl:sameAs when they are not in fact describing the same thing). DSNotify assists with the detection and repair of broken links by detecting structurally broken links and notifying the source data owner of changes to target datasets and can also be configured as a service that automatically forwards requests for moved target data resources to their new locations.

sparqlPUSH[21] is an interface that sits on top of a SPARQL endpoint, which allows the specification of a number of SPARQL queries into the dataset denoting which resources should be monitored and then uses the PubSubHubbub (Fitzpatrick, Slatkin, and Atkins, 2010) real-time Web protocol to broadcast updates to the RDF data store.

8.9   EVALUATING LINK QUALITY AND AVOIDING SEMANTIC SPAM

8.9.1   A WORD ON ACCURACY

It is an open question regarding how accurate or specific we can realistically expect Linked Data links to be. Many RDFS ontologies are quickly constructed, with the express aim of not trying to express the finer details of semantics, and owl:sameAs is frequently used to indicate any form of similarity or relatedness, without the author necessarily subscribing to or even fully understanding the implications of the Description Logics equivalence relation. Some sources are more trustworthy than others, and efforts under way into the expression of provenance, as discussed in Section 7.7, will no doubt help link architects to verify sources and improve link accuracy. However, it must be recognized that we are operating in a Web environment, and just as you do not assume that everything you read on the Web is true, you perhaps should not expect to believe the results returned from a Linked Data query that crosses multiple datasets. Future research might well look into how to increase the accuracy of query results; if, for example, the same information was repeated in multiple locations, this would lend credence to its veracity. In the meantime, systems must be designed that do not require or expect perfect data and treat information encoded in triples as claims rather than facts. In short, our advice is to be strict with your output and tolerant with your input.

8.9.2   Semantic Spam

Semantic spam is the term given to the misuse of Linked Data, or misrepresentation of information within Linked Data, to direct a semantic search engine or Semantic Web application to a spammer's data or Web site. When creating your own Linked Data and linking to other datasets, it is important to be aware of the tricks that could be used to insert false data. This will help you avoid using such techniques, however innocently, as semantic search engines will no doubt soon begin to detect and filter out datasets that employ these methods. This problem is in its infancy, but Ian Davis[22] has identified a number of semantic spam techniques. These include false labeling, identity assumption, false provenance, and manipulation of content negotiation.

In false labeling, well-regarded subject URIs are assigned an `rdfs:label` with spam content. Since `rdfs:labels` are often used for human-readable display when denoting an RDF resource, the spammer's message might well appear prominently in the Linked Data application in place of, say, Tim Berners-Lee's URI. Spam objects can also be inserted as the objects of triples involving other predicates commonly used to hold human-readable content, such as `isPrimaryTopicOf` or `rdfs:seeAlso`, or it is even possible for the predicate itself to be given a spam `rdfs:label`. Another direction of spam attack is identity assumption; `owl:sameAs` is used to misconnect a popular resource to a false resource that promotes the spam message. Since `owl:sameAs` is so widespread, many Linked Data applications use it for aggregating all triples about the subject together, so when querying for all data about `dbpedia:London`, say, you could find that spam triples are returned as well.

Another opportunity for spammers is false provenance; they attribute their message to a well-known and trustworthy person, for example, by saying

```
http://mereamaps.gov.me/PR/666 a bibo:Quote ;
    bibo:content 'I always drink at the Isis Tavern' ;
    dc:creator 'Tim Berners-Lee'.
```

This quotation could be displayed by a Linked Data application, along with its attribution, thus misleading consumers. A twist on this misattribution is to state the URI of a trusted individual or organization as the object of the triple instead of merely the text "Tim Berners-Lee."

Another trick outlined by Ian Davis is when useful Linked Data is supplied to the software agent, but spam messages are provided to humans, by manipulation of content negotiation processes. While a Linked Data application will make an HTTP request, using a Web browser aiming to supply human-readable information will send a different HTTP request, so the spam server can send different content to the two. This problem then means that it is particularly important for you as a well-regarded, nonspamming publisher to supply the same semantics in your Linked Data as human-readable content, as this is soon something that antispam filters will find a way to test for. (This recommendation is similar to the best practice guideline in the traditional Web of not hiding any data or links from the user.)

In addition to these "spam vectors" identified by Davis, you should bear in mind the issue with generating your own incoming links, a technique we suggested in

Section 8.4.6. If incoming links are also included in the Linked Dataset, a spammer could use this to misappropriate a third party's URIs and create triples with subjects from highly regarded Linked Data sources and spam objects. In the future, then, as search engines downgrade the importance of incoming links that are not resident in the external dataset, it will make the practice suggested in this section less useful.

Perhaps the most important lesson we can impart is that we should not be so arrogant we think we are better than our users or try and hide the data from them. To return to Ian Davis[23]: "Trust is a social problem and the best solution is one that involves people making informed judgements on the metadata they encounter. To make an effective evaluation they need to have the ability to view and explore metadata with as few barriers as possible."

8.9.3 LINKED DATA QUALITY

The issue of how to assess the quality of Linked Data sets has been addressed by a number of commentators.[24,25] First, one should assess the content. Is it logically consistent? Is the data accurate (are the facts correct?)? How frequently are updates made, and is the data current?

Second, one should judge the data model. Is it semantically correct? Is the data complete? Have a minimum of blank nodes been used? Are `rdf:resources` used rather than literals ("things not strings")? Have vocabularies been reused where possible? Are the URIs "cool" (Sauermann and Cyganiak, 2008)? Have the scope and purpose of the dataset and ontology been clearly stated? Does the dataset meet the stated scope and purpose, that is, is it complete and bounded? Have `rdf:labels` been used to make the data more comprehensible to human readers? What formats and access methods have been provided (for example, a SPARQL endpoint as well as an RDF dump)? Are there sufficient links to other datasets, particularly incoming links *that have been authored by third parties*, to indicate that this dataset is trusted?

Third, one can evaluate the provenance and usage of the data. Is it clearly and accurately attributed (can you tell where the data came from and who has edited it?)? What verification is possible? For example, is provenance information included, such as by the supply of a VoID dataset? Is the licensing clear? Will the data be maintained in the future? Is the publisher well known and authoritative?

While a reasoner can be run over an ontology, and SPARQL queries can be used as outlined in Section 8.3.5 to check data integrity, there are as yet no automated methods for answering the more subjective of these questions, particularly when assessing large datasets. Brand recognition and popularity assessment as represented by incoming links, as well as more explicit social media recommendations, are likely to be of most use in evaluating in Linked Data quality.

8.10 SUMMARY

This chapter has covered a number of areas surrounding the use and reuse of Linked Data. We have discussed some of the business models that have been suggested for exploiting the financial value of the data and covered the query mechanisms in the SPARQL RDF query language that not only allow information discovery but also assist with data validation. A large part of this chapter has been devoted to the important question of how to identify correspondences and relationships between datasets and how to craft appropriate linksets representing dataset integration through explicit triples whose subjects lie in one dataset and objects in another. Since a link may be semantically valid only in a certain context, we have also highlighted the need to be able to describe such contexts through mechanisms such as Named Graphs or Named Triples. Link maintenance and update protocols and processes are a technology in its infancy, but we highlighted some of the options available at the time of writing. Finally, we have addressed the issue of Linked Data quality and the pitfalls of semantic spam. As has become apparent in this chapter, Linked Data is a very young area, and as the volume of data published on the Linked Data Web grows, more techniques are being developed to tackle the problems that only emerge at scale and over time.

The nature of the development of Linked Data has necessarily meant that the development of methods and tools for using and linking Linked Data is less mature than those that aid the creation, databasing, and querying of Linked Data. As a result, whereas triple stores and SPARQL query engines are maturing and are now well supported, there is still a distinct lack of robust commercial-grade tool support for some of the newer areas. As the demands of industry grow, we can expect more products to come on the market that address the needs of the burgeoning Linked Data Web.

NOTES

1. http://www.chiefmartec.com/2010/01/7-business-models-for-linked-data.html
2. http://www.ldodds.com/blog/2010/01/thoughts-on-linked-data-business-models/.
3. http://data.nytimes.com
4. http://oauth.net
5. A draft standard for SPARQL 1.1 Update (Gearon, Passant, and Polleres, 2012) that has not yet been finalized allows for additional forms, including INSERT, DELETE, CREATE, and DROP, which allow RDF graphs to be altered in various ways. It is beyond the scope of this chapter to discuss further, but the interested reader can access the most up-to-date version of the forthcoming standard at http://www.w3.org/TR/sparql11-update/.
6. http://www.topquadrant.com/products/SPIN.html
7. http://clarkparsia.com/pellet/icv
8. http://sindice.com
9. http://swse.deri.org
10. http://ws.nju.edu.cn/falcons
11. http://sameas.org
12. http://www4.wiwiss.fu-berlin.de/bizer/silk/.
13. http://www4.wiwiss.fu-berlin.de/bizer/ldif/.
14. http://aksw.org/Projects/limes
15. http://keg.cs.tsinghua.edu.cn/project/RiMOM/.
16. http://www.w3.org/TR/2009/WD-sparql11-update-20091022/#t41
17. http://www.w3.org/TR/sparql11-http-rdf-update/.
18. http://www.ldodds.com/blog/2010/04/rdf-dataset-notifications/.
19. http://vocab.deri.ie/dady
20. http://docs.api.talis.com/getting-started/changesets

21. http://code.google.com/p/sparqlpush/.
22. http://blog.iandavis.com/2009/09/21/linked-data-spam-vectors/.
23. http://blog.iandavis.com/2007/11/21/is-the-semantic-web-destined-to-be-a-shadow/.
24. http://answers.semanticWeb.com/questions/1072/quality-indicators-for-linked-data-datasets
25. http://sourceforge.net/apps/mediawiki/trdf/index.php?title=Quality_Criteria_for_Linked_Data_sources

9 OWL

9.1 INTRODUCTION

In the previous chapters, we covered RDF (Resource Description Framework) and SPARQL, and explained how to organize Geographic Information into Linked Data. In Chapter 8, we highlighted some of the difficulties of linking different RDF datasets together due to the limited descriptions afforded by the RDF language. In this chapter, we introduce the OWL Web Ontology Language (Dean and Schreiber, 2004; World Wide Web Consortium [W3C], 2009), which offers a way of expressing more detailed knowledge about the domain of interest and of inferring new information from the set of statements we provide. This chapter sets out the main concepts of the OWL language and discusses the various options for software tool support. Chapter 10 takes you through the step-by-step process of using OWL to author an ontology, so that you will be in a position to use your ontologies to integrate data from different domains. This chapter does not attempt to provide a complete description of OWL since there are many other excellent publications that do this. Rather, it provides a flavor of the language and what it can and cannot be used for. The chapter ends with some examples of modeling best practice with some suggested design patterns.

OWL is not a single language but a whole family of related species. However, of these different species, one is almost universally used: OWL DL (Description Logic), so in this chapter we place our concentration here. To provide a more complete picture, for those interested, Appendix A describes the different OWL species and explains their differences and uses, and Appendix B provides details of the three syntaxes (Rabbit, Manchester Syntax, and RDF/XML [eXtensible Markup Language]) that we use in this chapter.

9.2 THE NATURE OF OWL

OWL, the Web Ontology Language, was first standardized as a recommendation of the W3C in 2004, with a more recent update, OWL 2, in 2009. Its purpose is to allow more expressive descriptions of knowledge than are possible with RDFS (RFD Schema), using formal logic to encode the semantic meaning in the statements.

In the Semantic Web "layer cake" shown in Figure 9.1, OWL is shown as a layer above RDFS, since it offers the knowledge modelers the opportunity to express more complex and detailed statements of their knowledge and perform more logical inference than is possible with RDFS. This does not necessarily mean that it is *better* to use OWL than RDFS as there are many cases for which an ontology allowing a simple set of triple statements will suffice. For example, it is likely to be more appropriate to use RDFS rather than OWL to describe the structure of a straightforward dataset, such as a set of names and ages of people. In this example, the class is Person

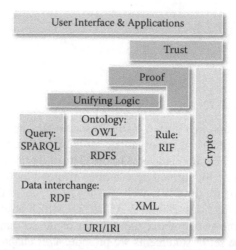

FIGURE 9.1 Semantic Web layer cake. (From http://www.w3.org/2007/03/layerCake. png. Copyright © 2007 World Wide Web Consortium, http://www.w3.org/ (Massachusetts Institute of Technology, http://www.csail.mit.edu/; European Research Consortium for Informatics and Mathematics, http://www. ercim.org/; Keio University, http://www.keio.ac.jp/). All rights reserved.

and has just two properties, "has_name" and "has_age." In RDFS, you can also create a property that indicates that one person is older than another. For example, Janet is older than John, and John is older than Sarah. But, what RDFS cannot do is enable a computer to make the logical step, based on the information given, that Janet is *also* older than Sarah. To carry out this reasoning, we need to rely on the OWL transitive restriction, which we apply to the property "older than." OWL is able to use this modified property to make the inference that if Janet is older than John and John is older than Sarah, Janet must also be older than Sarah. As we see further in this chapter, OWL can perform many more logical inferences than this, and care must be taken that the consequences of these reasoning steps are still accurate statements. The decision of whether to use RDFS or OWL should err toward simplicity: stick to RDFS if possible. However, say you wanted to pass your People's Ages dataset on to someone else, who might need to know whether the names were firstname/surname structure or vice versa and on what date the ages were calculated or have various other questions about the dataset that need to be answered before being reused or integrated with personal data. In this case, it will be worth investing the extra time needed to express this more complex ontology in OWL.

The basic language constructs of OWL are the same as those we have encountered in RDFS: classes, properties, individuals, and datatypes. An OWL document is similarly made up of statements or assertions in the form: "Subject Predicate Object."

9.1.1 Differences between OWL and Object-Oriented Languages

For those used to object-oriented (OO) design and programming methods, it is easy to think that OWL is an OO language or at least OO like. However, there are very significant differences; the implications of these differences are not always obvious, so it is worth spending a little time explaining them by looking at how OWL (and RDFS) classes differ from OO classes and properties.

In OWL, classes are regarded as *sets* of instances, which are known as "individuals" in OWL terminology. In comparison, OO languages use classes to represent the *type* of the instances. This means that it is often helpful to picture OWL statements in terms of Venn diagrams of overlapping sets to understand where the groups of individuals sit. To make the difference clear, in OWL a class is a collection or set of individuals: The class Car would contain all the individuals considered to be cars. By contrast, in OO language, a class defines the conceptual type, that is, the data structure that can represent information about cars with their allowed behaviors implemented by functions. OWL individuals can belong to multiple classes, whereas in the OO paradigm, barring multiple inheritance, which is generally discouraged or not supported in many OO languages, each instance can only have one class type.

In OWL, classes can be created and changed at run time, and OWL is based on the open world assumption, whereas OO language is based on the closed world assumption. Recall the explanation that we gave in Chapters 1 and 4 about the open world assumption: Just because we have not said so does not mean an assertion is not true. We cannot guarantee that we have discovered all information about the system, and statements about knowledge that are not stated or inferred from statements are considered unknown rather than false. In some ways, this can be thought of as an assumption of "innocent until proven guilty": Even if there is not enough information to prove a statement true, it still cannot simply be assumed to be false. So, an OWL class specifies the mandatory properties that describe the essence of that class, whereas an OO class defines all the possible properties the class can have. Furthermore, even if a property is mandatory in OWL, it is not mandatory to know the value of that property for any individual. In OWL, properties are stand-alone entities that exist without specific classes, unlike OO language, for which properties are defined locally to a class.

Another major difference is related to their purpose. OO languages are designed to implement processes; hence, OO classes encode their meaning and behavior through functions and methods. In contrast, OWL is designed to provide description, and there is no code to specify the actions of an object. Instead, classes make their meaning explicit solely in terms of OWL statements: They *are*, but they do not *do*.

OWL classes and individuals can be linked to from anywhere on the Web, so there is no encapsulation using public or private access as in OO languages. The domain models for OO software, often encoded in UML, tend to be used internally to the organization only and are decoupled from the actual code, whereas all OWL domain models can be shared on the Web. Similarly for individuals, in OWL and RDF individuals can be reused, whereas OO instances are internal to the executing program and cannot be addressed from outside the runtime environment.

The main implication that can be drawn from these differences is that you cannot think about an OWL class in the same way that you would an OO class; they are very different beasts. OWL classes are sets of individuals, each sharing some common properties but each also unique and varying from its companions. You will find that it is easiest to understand OWL classes by thinking about the individuals in the class set.

9.1.2 Syntaxes

Since an OWL ontology is an RDF graph, it can be written using many different syntaxes. We have already met the RDF/XML syntax in Chapter 5, and the RDF constructs we are familiar with also form part of the RDF/XML syntax for OWL. In addition, there are tags such as `owl:Class` that are used as part of the OWL RDF/XML syntax. This is the only syntax that all OWL tools are required to support, so you are likely to come across OWL ontologies formatted this way.

The Manchester Syntax (Horridge and Patel-Schneider, 2009) was introduced in the OWL 2 standard with the aim of making it easier to read and write ontologies. It is more compact and easier to understand than RDF/XML, so we use it in our OWL examples in this book.

For completeness, we also mention the OWL/XML syntax (Motik, Parsia, and Patel-Schneider, 2009), which is easier to parse than RDF/XML and unlike the latter, can be processed and queried using off-the-shelf XML tools like XSLT and XQuery. However, like RDF/XML, it is still quite verbose, so we stick to the Manchester Syntax for the rest of this chapter.

Although not part of the formal OWL standard, several controlled natural language (CNL) syntaxes have been proposed for authoring OWL ontologies. These include Attempto Controlled English (ACE) (Kaljurand and Fuchs, 2007); the Sydney OWL Syntax (Cregan, Schwitter, and Meyer, 2007); and Rabbit (Hart, Johnson, and Dolbear, 2008). Their purpose is to make it easier for experts in the particular domain of knowledge—flood risk management, oncology, patent law, or whatever it may be—to capture their own knowledge as easily and naturally as possible, using sentences that are close to normal English, while maintaining the unambiguity necessitated by OWL. The main differences between these three is that ACE and the Sydney OWL Syntax both started life as more complex controlled languages, based on first-order logic (the latter as the language PENG), and were later reduced in scope to conform to OWL DL, while Rabbit was developed by domain experts looking to author, and understand, their own ontologies. In this book, we include Rabbit sentences as an explanation of the OWL DL examples. Appendix B summarizes the primary constructs of Rabbit and their corresponding OWL DL axioms in Manchester and OWL/XML syntaxes.

9.3 OWL LANGUAGE ELEMENTS

9.3.1 Ontology-Level Constructs

An OWL ontology consists of a set of statements known as "axioms" that are usually preceded with a set of namespace declarations. As with RDFS, these axioms consist of triples with the structure <subject> <predicate> <object>, and like RDFS, OWL implements these through classes, properties, individuals, and values.

The namespace declarations are there to identify the ontologies that are imported and to denote the current ontology's own prefix, which is then used in the rest of the ontology as a shortcut. It can be useful to introduce this prefix so that if the domain

URL (Uniform Resource Locator) needs to be changed, only one change need be made rather than strings needing to be changed throughout the document.

There are a number of standard namespaces for ontologies:

rdf	http://www.w3.org/1999/02/22-rdf-syntax-ns#
rdfs	http://www.w3.org/2000/01/rdf-schema#
xsd	http://www.w3.org/2001/XMLSchema#
owl	http://www.w3.org/2002/07/owl#

Any ontology editor will provide these for you, but it is worth being able to recognize them. To use definitions originally coined in other ontologies, you may wish to import them into your ontology. Importing an ontology means that you take on board, and agree with, *all* the statements in the third-party ontology, so you may wish to segment it first and only take the subset of statements with which you agree. This will also make the ontology smaller.

The following is an example of how Merea Maps' ontology might begin:

```
1. Prefix: <http://mereamaps.gov.me/topt/>
2. Prefix: dc: <http://purl.org/dc/elements/1.1/>
3. Prefix: rabbit: <http://www.ordnancesurvey.co.uk/ontology/Rabbit/
v1.0/Rabbit.owl/>
4. Ontology: <http://mereamaps.gov.me/topo> <http://mereamaps.gov.me/
topo-v1>
5. Import: <http://www.ordnancesurvey.co.uk/ontology/Rabbit/v1.0/
Rabbit.owl>
6. Import: <http://purl.org/dc/elements/1.1/>
7. Annotations:  dc:rights "Merea Maps 2011",
8.                dc:title "Topography",
9.                rabbit: purpose "To describe the administrative
geography of Merea and the topological relationships between them."
10.      rabbit: scope "All levels of administrative area that occur in
Merea, their sizes, point locations, the topological relationships
between areas of the same type. Authorities that administer the regions
are not included, nor are the spatial footprints of the regions."
11.      owl:versionInfo "v1"
12.AnnotationProperty: dc:rights
13.AnnotationProperty: dc:title
14.AnnotationProperty: rabbit:purpose
15.AnnotationProperty: rabbit:scope
16.AnnotationProperty: owl:versionInfo
```

The line numbers in this code are for reference only and not part of the actual ontology. As we can see, the ontology begins by stating the prefixes used: first, the main ontology, which has no prefix; then the prefix dc for the Dublin Core ontology, which is a well-known standard ontology that provides annotation properties like title, rights, and coverage; and the prefix rabbit for the CNL ontology that includes annotation properties linked to the ontology authoring method described in Chapter 10, for example, to include annotations on the scope and purpose of the ontology. Line 4 states the URI for the current ontology and optionally a Version URI. It is common practice to number the versions of an ontology, but the most up-to-date version will have an unversioned URI. In this example, we are looking at the most up-to-date version of the ontology, which is also version 1, hence the two separate URIs stated on line 4. Lines 5 and 6 state that we are importing the Dublin Core and Rabbit ontologies into our Administrative Geography ontology, and lines 7–11 are annotations that apply to the whole ontology. It is also possible to annotate a single entity (i.e., a class or property) or an axiom. Frequently, the `rdfs:label` annotation property is used to annotate each class and property to provide a "human-friendly" version of the class or property name. For example, while the class name might be `LicensedEstablishment` as one single word, it would also be labeled with "Licensed Establishment"—allowing spaces, accents, numbers, or other symbols that are forbidden in OWL class or property names. Good ontology editing tools are then able to use these labels to provide a more readable view on the ontology, and an author can use the labels instead of the official names to handle the concepts.

The annotations on lines 7 and 8 use Dublin Core annotation properties to state who owns the copyright on the ontology and the ontology title, respectively. Lines 9 and 10 state the purpose and scope of the ontology, which are described more in Chapter 10. Line 11 gives the version number of the ontology. Lines 12–16 state the annotation properties that are used in this ontology.

Ontology annotations provide metadata to help a user analyze and compare the ontology to others for mapping and management purposes. They do not provide any semantic meaning to the ontology and are not included in the reasoning process. As well as `owl:versionInfo`, OWL DL provides some other constructs for annotation purposes: `owl:backwardCompatibleWith`, `owl:incompatibleWith`, `owl:deprecated`, and `owl:priorVersion`. Built-in annotation properties from RDFS, which we discussed in Chapter 7, such as `rdfs:label`, `rdfs:comment`, `rdfs:seeAlso`, and `rdfs:isDefinedBy`, can also be used.

9.3.2 CLASSES

The root class of an OWL ontology is `owl:Thing`, a predefined, universal (set) class that includes all individuals. Every concept we talk about in the ontology inherits from this root. Since OWL also incorporates RDFS, an `rdfs:Resource` is a kind of `owl:Thing`, an `rdfs:Class` is a kind of `rdfs:Resource` (as we already know), and an `owl:Class` is a kind of `rdfs:Class`. OWL also predefines the empty class `owl:Nothing`, which has no members. If the results of your reasoning end up putting any of your individuals into the `owl:Nothing` class, you have gone wrong somewhere. The set of classes and statements about classes is known as the Terminological Box or Tbox. The set of instances or, to use OWL terminology, Individuals and facts about them is known as the Assertion Box or Abox.

To make the Rabbit statement "Pub is a Concept." in OWL, we would need to say

```
Class:Pub
SubClassOf: owl:Thing
```

Following on from this, we can also make the statement that one class is a subclass of another, such as "Every Freehouse is a kind of Pub".

```
Class:Freehouse
SubClassOf:Pub
```

These statements are part of the Tbox as they are statements or axioms about a class.

In OWL, a class can also be constructed by enumerating all its allowable instances, using the OWL property `owl:oneOf`. For example, we can describe beer by listing the types of beer[1]:

Every Beer is one of Mild, Pale Ale, Brown Beer, Barley Wine, Old Ale, Porter, Stout or Lager.	`Class:Beer`
	`EquivalentTo: {Mild, PaleAle,`
	`  BrownBeer, BarleyWine, OldAle,`
	`  Porter, Stout, Lager}`

The `owl:oneOf` construct can also be used to define a range of data values, known as an enumerated datatype. For example, the Merea Land Management Agency could specify that the datatype property `hasTaxCode` can only take one of a fixed number of numeric tax codes. As you can see from the rather contrived example, this is likely to be used only rarely in reality.

Two classes can be denoted as equivalent through the use of `owl:equivalentClass`; for example, Gas Station is equivalent to Petrol Station. This is quite a strong logical assertion because it means that every individual in one class is also an individual in the other class, and all the statements made about each class apply to the other class.

Gas Station and Petrol Station are Equivalent.	`Class: GasStation` `EquivalentTo: PetrolStation`

Conversely, `owl:disjointWith` states that no member of one class is a member of the other. For example, Field and Barn are mutually exclusive:

Field and Barn are mutually exclusive.	`Class: Field` `  disjointWith: Barn`

If you think about it, there are thousands of examples for which this is true (although be careful of the edge cases, where it may not always be true, such as River and Stream). However, the decision of whether to clutter up your ontology with such statements should be made based on the stated purpose of the ontology. Including many disjoint statements can slow inference and make the ontology less readable, more error prone, and potentially less useful for data integration purposes. Disjoint statements should only be included when they are necessary for the purpose of the ontology: to answer a specific competency question, or to make it clear when something really is not allowed, for example, if the tax rules in Merea are such that a property must be either a business premises or a residential one. In this case, you would want to make these two classes disjoint to flag any discrepancies in the data if a premises had been misclassified as both.

If several classes are mutually disjoint, OWL 2 has introduced an additional property owl:disjointClasses, which avoids the need to use owl:disjointWith between every pair in the group of mutually disjoint classes, making such statements of disjointness more concise, easier to read, and less error prone. Another OWL 2 property, owl:disjointUnion, allows the specification of all possible disjoint subclasses of a class. Using the tax example, the following states that a Premises is exclusively either a BusinessPremises or a ResidentialPremises and cannot be both of them (at least from the perspective of Merean taxation):

```
Business Premises and Residential     Class: Premises
  Premises are mutually exclusive.     DisjointUnion: BusinessPremises,
Premises is a kind of                      ResidentialPremises
Business Premises or
Residential Premises.²
```

Disjoint Classes and Disjoint Union are often called "syntactic sugar" as their meanings can be constructed using multiple Disjoint With and Union statements. They are just offered as a shorthand alternative rather than to provide new expressivity to the language.

9.3.3 INDIVIDUALS

An "Individual" is the name given in OWL to instances or members of a class, and statements about individuals are contained in the Abox. We can state that "The Isis Tavern is a Pub." in the following way:

```
The Isis Tavern is a Pub.             Individual:
The Isis Tavern has name                http://mereamaps.gov.me/topo/0012
 "The Isis Tavern".                   Types: Pub
The Isis Tavern and Frog and          Facts: hasName value "The Isis
 Frigate are different things.           Tavern"^^xsd:string
                                      DifferentFrom:
                                        http://mereamaps.gov.me/topo/0011
```

This says that the information resource controlled by Merea Maps with URI http://id mereamaps.gov.me/topo/0012 is a member of the class Pub and has the name (as a string value) "The Isis Tavern," using the Manchester Syntax keyword "Facts," and that it is a different individual from the pub represented by the URI http://id.mereamaps.gov.me/topo/0011.

The construct owl:AllDifferent states that in a list of individuals, all are mutually distinct from each other and can be used as a shortcut instead of making separate pairwise statements using owl:differentFrom. The Manchester Syntax uses the keyword DifferentIndividuals to denote this:

```
DifferentIndividuals: AshFleetFarm, HomeFarm, IsisFarm
```

To state that the individual is the same individual as another, OWL has the keyword owl:sameAs. In reasoning terms, this means that *all* information about one individual also applies to the other individual.

The Frog and Frigate and The Frog and Frigate [Big Breweries] are the same thing.	Individual:
	  http://mereamaps.gov.me/topo/0012
	sameAs:
	  http://bigBreweries.com/00012345

We have already seen how Linked Data frequently uses the owl:sameAs property to marry different datasets, but caution is needed as it only applies when the information on both sides is equally valid for both individuals. If not, it is better to choose another relationship to link the two, such as "related to," or a property that describes a more specific semantic relationship as appropriate: "belongs to," "owns," "part of," and so on.

9.3.4 VALUE CONSTRAINTS

Value constraints are axioms that restrict the values that can be assigned to a class. An example of the simplest form of value constraint is

Every Pub sells Beer.	Class:Pub SubClassOf: owl:Thing that sells some Beer

This states that a pub must sell some beer, that is, at least one individual from the set of Beers. In Manchester Syntax, the word *some* is explicitly used, whereas in Rabbit it is implied. The RDF/XML version of this term, which you may also come across, is owl:someValuesFrom. The use of "some" (whether explicit or implied) is known as an existential quantifier. This comes from the logic symbol $\exists$, which should be read as "there exists," so the statement we are making is that for every Pub there exists at least one Beer that is sold there.

It is important to understand exactly what this statement is doing as it helps to reinforce the way of thinking that is embodied in OWL. OWL and RDFS are set based: Classes are effectively sets of things. So, the class Pub is a subclass (or subset) of an anonymous class known as a Restriction, as it is defined by adding a restriction on the property "sells" so that it is restricted to the range of at least one Beer. The anonymous class only requires that at least Beer is sold. Other drinks could also be sold, which is what we want from a Pub. A restriction on the range of a property is called a "value constraint" as it restricts which values the property can take.

Things get a little more complicated with this second example:

Every Beer is sold only in a Licensed Establishment or nothing.	Class:Beer Subclassof: owl:Thing that isSoldIn only LicensedEstablishment

The second kind of value constraint uses "only" to represent the $\forall$ logic symbol, also known as the universal quantifier. The RDF/XML term for this that you may come across is `owl:allValuesFrom`. This value constraint is much more restrictive than "some" as it creates an anonymous class that contains *only* things that are sold in a Licensed Establishment *and nothing else*. Essentially, this says that beer can only be sold somewhere that has a license to sell alcohol (a licensed establishment); it cannot be sold anywhere else. So, you can see how careful we have to be with applying this restriction as in many cases it would convey incorrect knowledge. For example, applying the universal quantifier (only) to our first example would make the statement "Every Pub sells only Beer or nothing" and would mean that there are not any Pubs that sold other beverages, which would be a tragedy for cider drinkers.

The other important point to note about "only" is that the anonymous class could be empty (whereas for Some we know it must contain at least one member). This means that what we are really saying in the second example is that "Beer can only be sold in Licensed Establishments and nowhere else, but it might not be sold at all" (a tragedy for beer drinkers).

An ontology pattern that is often seen is when both types of restriction are used. This changes the statement to "Beer is sold in Licensed Establishments only, and nowhere else, *and it has to be sold*." The existential quantifier "some" adds the italicized part of the statement. This is known as a "closure axiom":

```
Every Beer is only sold in        Class: Beer
  Licensed Establishments.          SubClassOf: owl:Thing that isSoldIn
                                       onlysome LicensedEstablishment
```

The Rabbit statement is simpler in that it assumes that "only" implies the closure axiom as this is the most likely meaning in natural language. In the OWL file, you may notice that your ontology editor separates the `onlysome` keyword out into two statements:

```
Class: Beer
SubClassOf: isSoldIn only LicensedEstablishment and isSoldIn some
LicensedEstablishment³
```

Another value constraint is `owl:hasValue`, which corresponds directly to the CNL term "has value" and the Manchester Syntax keyword `value`. It is used to specify a particular value that a property must take when used with a certain subject class or individual. As we saw previously:

```
The Isis Tavern has name          Individual:
  "The Isis Tavern".                 http://mereamaps.gov.me/topo/0012
                                    Types: Pub
                                    Facts: hasName value "The Isis Tavern"
```

9.3.5   CARDINALITY CONSTRAINTS

As well as value constraints restricting properties, OWL specifies cardinality constraints: the maximum, minimum, or exact number of values that a property can take when applied to a particular concept. These are often known as Qualified Cardinality Restrictions. (The designation "qualified" means that the property restriction only applies when the property value is a particular concept, rather than applying to the property globally.) For example, we can describe a semidetached house as a kind of house that is attached to exactly one other house.[4]

```
Every Semi Detached House is        Class: SemiDetachedHouse
attached to exactly 1 Semi          SubClassof: House that
Detached House.                        (isAttatchedTo exactly 1
                                       SemiDetachedHouse
```

Similarly, "max" or "min" (in Rabbit "at most," "at least") can be used to denote the maximum or minimum value that the property can take.

9.3.6 INTERSECTION AND UNION

As we know, OWL is set based, and two commonly used concepts are Union (where we are talking about any individuals that belong to either one or both overlapping or possibly nonoverlapping sets) and Intersection (where we are interested in only those individuals that are members of both sets at the same time). In the OWL/XML syntax, these are known as owl:UnionOf and owl:IntersectionOf, respectively. These correspond to the logical OR and AND, but they do not quite correspond to how we understand the words *or* and *and* in straightforward English. Rabbit enables the reader to clearly differentiate between an Exclusive OR, "you may have tea or coffee" (but not both), and the Inclusive OR, "A farm can have land that comprises arable fields and/or pasture and/or meadow," meaning that it can have any combination. This last form is always preceded by "one or more of" in Rabbit. So, the Rabbit and Manchester versions of the last statement would be

```
Every Farm has land one or more     Class: Farm
of Arable Fields or Pasture or      subClassOf: hasLand some
Meadow.[5]                            ArableField or Pasture or Meadow
```

Figure 9.2 visualizes this statement in terms of sets. Note that "or" means "one or the other or both" (so the farm in the example contains some land that is arable land, pasture land, or meadow, or land that is a mixture of two or three of them). As we can see from the members of the sets in the Venn diagram of Figure 9.2, there are a number of farmland instances that lie within the union of the three sets. Conversely, however, an intersection restriction would mean that every option mentioned must be true. So, if we had said

```
Class: Farm
subClassOf: contains some ArableLand and PastureLand and Meadow
```

we would be stating that the land was Arable and Pasture and Meadow all at the same time, that is, referring to the center of the Venn diagram where all three circles intersect, which is empty. This example illustrates the beginner's mistake of using the word *and* in English to refer to a logical OR. In fact, Rabbit does not allow the use of "and" in order to prevent this very common mistake, so intersection must be stated through

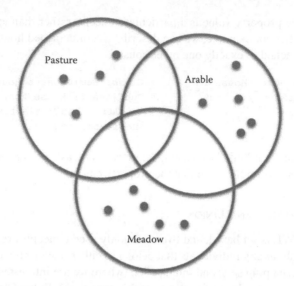

FIGURE 9.2 Venn diagram of farmland types. Spots represent members of the three sets Arable, Pasture, and Meadow.

FIGURE 9.3 Venn diagram of organizations' responsibilities.

separate Rabbit statements as shown in the Land Management Agency example that follows. Whenever you encounter the keyword "or" in OWL, you should understand it to mean "either one or other or both," and the keyword "and" in OWL refers to when the thing is a member of *both* sets. For example, the Land Management Agency has responsibility for Land Registration and Property Valuation.

```
Land Management Agency has          Individual: LandManagementAgency
 responsibility for Land Registration.  Facts: hasResponsibilityFor
Land Management Agency has           some LandRegistration and
 responsibility for Property Valuation.  hasResponsibilityFor some
                                     PropertyValuation
```

It sometimes flows better to use the word *that* when linking two classes, so both the Rabbit and Manchester Syntaxes also allow "that" as a keyword instead of "and" to aid the reader, although again the use of "that" within Rabbit is more restrictive.

9.3 PROPERTIES

OWL uses the standard RDF properties, with accompanying domain and range restrictions, and as with RDF, subproperty relationships are also possible. These can be represented in the Rabbit and Manchester Syntaxes as in the following examples:

```
"sells" is a Relationship that takes a          ObjectProperty: sells
concept as an object.                            Domain: Pub
The relation "sells can only have Pub as a       Range: Beer
subject.
The relation "sells" can only have Beer
as an object.
```

And a datatype property can be denoted as

```
"has nick name" is a Relationship that           DataProperty: hasNickName
takes a value as an object.                       Domain: Place
The relation "has nick name" can only have        Range: xsd:string
a String as a Value.
```

Note that the convention for naming properties in Manchester Syntax (and syntaxes other than Rabbit, which uses lowercase throughout) is to use camel case, that is, the first word is lowercase and the first letter of subsequent words is capitalized. While domain and range restrictions on properties are inherited by their subproperties, this is not the case for other property characteristics like symmetry or transitivity, which we discuss in a separate section. Most property constraints are global, that is, they apply for every class that uses that property. In contrast, a "local" property constraint will only apply to the property's use with a particular class. Unless otherwise indicated, the property constraints we discuss are all globally applicable.

9.3.7 EQUIVALENT PROPERTY

`owl:equivalentProperty` states that two properties have the same property values for a given individual. It is important to note that this is not the same as stating that the properties mean the same thing. (As an aside, to do that you would have to use the `owl:sameAs` construct, which would mean treating the classes as individuals themselves, which is only allowed in OWL Full or OWL 2.0. For more details, see Appendix A.)

9.3.8 INVERSE OF

The `owl:inverseOf` construct applies to two properties that link the same classes while reversing the subject and object. For example,

```
The relationship "owns" is the complement         ObjectProperty: owns
of "owned by".                                     InverseOf: owned_by
Merea owns MereaMaps.                              Individual: Merea[6]
                                                   Facts: owns MereaMaps
```

If we know that Merea owns Merea Maps, then we can use the inverse property relationship between "owns" and "owned by" to infer that Merea Maps is owned by Merea.

9.3.9 SYMMETRY

The relationship "next to" is symmetric.	ObjectProperty: nextTo Characteristics: Symmetric.

If a property such as "next to" is symmetric, then this means that if A is next to B, then we can also infer that B is next to A.

The relationship "is larger than" is asymmetric.	ObjectProperty: isLargerThan Characteristics: Asymmetric.

Conversely, an asymmetric property means that if A is related to B, then B cannot be related to A. For example, the relationship "is larger than" is asymmetric, as if one thing is larger than another, the second thing cannot ever be larger than the first.

9.3.10 TRANSITIVITY

The characteristic of transitivity can be seen in the following example with the property "connected to": If roadStretch1 is connected to roadStretch2 and roadStretch2 is connected to roadStretch3, then roadStretch1 is also connected to roadStretch3.

The relationship "connected to" is transitive.	ObjectProperty: connectTo Characteristics: Transitive.

As we discussed in Chapter 10, transitivity needs to be treated with care but can be very useful.

9.3.11 FUNCTIONAL, INVERSE FUNCTIONAL, AND KEY PROPERTIES

A functional property is many to one: a property that can only have one unique value as the object for each individual using that property. For example, "is capital city of" can only take one value as an object, as a city can only be the capital city of one country. "is capital city of" is also an example of an inverse functional property (one to many), that is, one that can only have one value as the subject: Only one city can be the capital city of a country. In Rabbit CNL and Manchester OWL Syntax, this is stated as

The relationship "is capital city of" can only have one object. The relationship "is capital city of" can only have one subject.	ObjectProperty: isCapitalCityOf Characteristics: InverseFunctional ObjectProperty: hasCapitalCity Characteristics: Functional

This can be useful for making inferences. For example, given

```
Medina is the capital city of Merea.    Individual: Medina
Meddy is the capital city of Merea.     Facts: isCapitalCityOf Merea
                                        Individual: Meddy
                                        Facts: isCapitalCityOf Merea
```

Since we know that isCapitalCityOf can only have one object, then the two objects we have been given, Medina and Meddy, must be the same thing:

```
Medina and Meddy are the same thing.    Individual: Medina
                                        SameAs: Meddy
```

The key property in OWL 2 (`owl:hasKey`) is akin to the idea of a primary key in a relational database, allowing you to provide a separate identifier (as well as the URI) for each individual. In practice, this is mostly used to store the primary key from the database when converting relational data to Linked Data in RDF. As with relational data, a key can be made up of a set of properties. In OWL, a key property is not automatically a functional property, that is, the key property may take more than one value, although you can certainly state that it is functional, if desired. A statement using the `owl:hasKey` property says that each named instance of a class is uniquely identified by a (data or object) property or set of properties. That is, if two named instances of the class have the same value for their key property, it can be inferred that they are the same individual. There is no equivalent statement in Rabbit.

```
Class: TaxableProperty
HasKey: hasTaxationNumber
```

This says that each taxable property is uniquely identified by a taxation number. If we then use Ash Fleet Farm as an example of a taxable property and state that it has the taxation number 12345,

```
Individual: AshFleetFarm
Facts: hasTaxationNumber "12345"
```

and if we then come across another individual with the taxation number 12345, we can infer that it is Ash Fleet Farm. You can see that this is similar to the impact of an inverse functional property except that it only applies to individuals that are explicitly named, and we cannot draw the inference that every individual that has a taxation number belongs to the class TaxableProperty.

9.3.12 Reflexivity

Reflexivity means that the relationship holds between a class and itself. For example, the property part_of is sometimes seen as reflexive.[7] Therefore, for *any* class that uses the part_of relationship, it will be possible to say that the class is part of itself:

```
Everything is part of itself.              ObjectProperty: part_of
                                           Characteristics: Reflexive
```

Note that the characteristic of reflexivity belongs to the *property*, not to any particular class. Alternatively, a property can be **irreflexive**, meaning that no individual can be related to itself by such a property. Previously, we used the example of a Semidetached House as follows:

```
Every Semi Detached House is attached        Class: SemiDetachedHouse
  to exactly 1 Semi Detached House.          SubClassof: House that
                                               (isAttatchedTo exactly 1
                                               SemiDetachedHouse
```

And we noted that this axiom allows a semidetached house to be attached to itself, which is rather unhelpful. By making the property isAttachedTo irreflexive, then we now ensure a semidetached house can only be attached to one other and different semidetached house.

`owl:objectHasSelf` is a construct used to describe the characteristic of **local reflexivity**, that is, reflexivity for individuals. It can be applied to the class of objects (individuals) that are related to themselves via the given object property. It is useful when reflexivity does not hold for the property globally, but for a certain number of classes, the property will still be reflexive. For example, an autoregulating ecological process is one that regulates itself. So, local reflexivity could be applied to the property "regulates" if the subject is an `AutoRegulatingEcologicalProcess`.

```
Class: AutoRegulatingProcess
SubClassOf: regulates some Self
```

9.3.13 NEGATIVE ASSERTIONS

Due to the open world nature of OWL, if we do not state a fact, it could still be true. So if, say, we do not know whether Merea Maps is a subsidiary of the Merea Land Management Agency, we need say nothing. However, if we definitely know that it *is not* a subsidiary, and this is important for some reason, we can state this as follows:

```
Merea Maps is not a subsidiary of          Individual: MereaMaps
  Merea Land and Property.                  Facts: not isSubsidiaryOf
                                             {MereaLandAndProperty}
```

OWL 2 also introduces something called a Negative Property Assertion, which does not have a corresponding Rabbit construct:

```
Individual: MereaMaps
NegativeObjectPropertyAssertion: isSubsidiaryOf MereaLandAndProperty
```

Furthermore, OWL 2 also supports a Negative Data Property, which can be used to model statements such as "MereaMaps does not have 100 employees."

9.3.14   PROPERTY CHAINS AND GENERAL CONCEPT INCLUSION AXIOMS

Another useful feature introduced in OWL 2 is the property chain, which is also known (for certain reasoners and so you may come across this term) as a *complex role inclusion axiom*. A property chain defines a property as a chain of object properties. The clearest example of this is in the domain of families: An aunt is the sister of one's parent. So, the property `hasAunt` can be defined as the property chain `hasParent o hasSister` (the DL symbol *o* is often used to denote a link in a property chain).

Everything that has a Parent that has a Sister will also have that as an Aunt.	ObjectProperty: hasAunt subPropertyChain: hasParent o hasSister
Everything that has a Part that contains some Thing will also contain that Thing.	ObjectProperty: contains subPropertyChain: hasPart o contains

You may also come across the term *general concept inclusion* (GCI) axiom, which is a more specific version of the complex role inclusion axiom, involving a concept associated with the property chain, for example, "Everything that has a Part that contains some Water will also contain some Water".

A GCI occurs when C is a subclass of D, where C is any general concept and C and D can be complex (e.g., anonymous classes). A GCI can be thought of as any statement that has more than just a single class name as the subject. The more of them there are in your ontology, the longer it will take to reason over. This is also true for the property chains. However, these types of axiom are often the "glue" that holds the ontology together and helps you achieve the right answers to your competency questions. If, when you query your ontology with the competency question "test set," you do not see the answer you are expecting, consider whether you need to add some complex role inclusion axioms or GCIs to create the correct chain of reasoning.

9.4   TOOLS FOR AUTHORING

To manage the creation and maintenance of an ontology, tool support is essential. There are two main types of ontology tool—editors and reasoners—although some products combine the functions of both. An ontology editor is used to create and edit an ontology, while a reasoner allows queries to be asked of the ontology and determines the implicit knowledge that is a consequence of the ontology's statements. The most widely known tool for creating OWL ontologies, as well as RDF/XML, is the free, open source editor Protégé,[8] produced at Stanford University (Musen, 1988). Protégé 3.x is a frames-based editor, that is, one that organizes an ontology into a set of classes, slots (for properties and relationships), and instances and has been evolving over a number of years. The more commonly used version is Protégé OWL (Protégé version 4), an extension of the Protégé editor built at the University of Manchester. This allows OWL and RDF ontologies to be loaded, saved, edited, and visualized. In addition, rules encoded in the Semantic Web Rule Language (SWRL) (Horrocks et al., 2004) can be added. There are several useful plug-ins, for example, the reasoners HermiT,[9] FaCT++,[10] and Pellet,[11] and plug-ins for editing in the CNLs Rabbit[12] (called the ROO tool) and AceView.[13] There is also an online version called WebProtege.[14] Other editors are available, such as the open source NeOn-Toolkit,[15] which is a plug-in to the Java Integrated Development Environment Eclipse, the domain-specific Snow Owl for clinical terminologies,[16] the online cloud-based Knoodl,[17] and the commercial TopBraid Composer[18] and FluentEditor[19] (the latter is CNL based). Which of these is selected for use will depend on your budget, project size, and requirements; however, the main influencing factors will be

- Commercial versus open source: While the commercial editors' price tag will come with product support and more robustness, there are strong communities surrounding the main ontology editors Protégé and NeOn, so questions can be answered on their forums, and there are a larger number of useful plug-ins.
- RDFS versus OWL: If the main aim is to create a Linked Data set, a tool that is limited to creating RDFS ontologies will be sufficient as priority must be given to the management of large datasets. For more details of such tools, see Chapter 7. Instead, if the aim of the project is to author highly descriptive ontologies or to integrate ontologies together, one of the ontology editing tools mentioned that support OWL 2 should be used.
- Reasoning, rules, and queries.

The OWL 2 reasoners FaCT++, HermiT, and Pellet are all offered as stand-alone or as plug-ins to Protégé 4. RacerPro[20] is a commercial reasoning system that is available as a stand-alone product, with a visualization tool RacerPorter to manage the knowledge bases. There are other reasoners that support subsets of OWL 2,[21] for example, ELK[22] and CEL,[23] which support OWL 2 EL; QuOnto, which supports OWL QL; and Oracle 11g relational database management system, which supports OWL RL. If you know your ontology conforms to one of the OWL profiles, it is generally better to choose a reasoner that corresponds to its logical complexity as they are faster and lighter than the reasoners that cover the whole gamut of OWL 2 logic. If you cannot decide, there is always TrOWL,[24] which is an interface to a number of reasoners and offers EL ("TrOWL REL") and QL ("TrOWL Quill") or uses FaCT++, HermiT, or Pellet for full DL reasoning. HermiT, Pellet, and RacerPro also provide support for SWRL rules and allow SPARQL or SPARQL-DL (a subset of SPARQL that supports OWL DL-based semantics) queries.

- CNL support. As mentioned, Protégé has some plug-ins to enable authoring in a CNL, for example, ROO and CloNE (Funk et al., 2007), and the commercial tool Fluent Editor has integral support for its CNL. All others allow annotation of the corresponding CNL statement using the `rdfs:comment` construct. If a domain expert, unfamiliar with OWL, is responsible for authoring the ontology directly, then we would strongly advise considering the use of an editing tool that supports CNL.

- Ontology development life-cycle support. Different tools offer different levels of support for the various stages of ontology development, including requirement gathering; versioning; issue tracking; collaboration (Collaborative Protégé, Knoodl); merging (PROMPT plug-in for Protégé); visualization (Altova SemanticWorks,[25] OWLViz plug-in for Protégé); and debugging (Protégé 4), among others.

9.5 SUMMARY

This chapter has worked through the main concepts in the Web Ontology Language OWL, namely, metadata annotations, Classes, Properties, and Individuals. We have also discussed the primary tools available for authoring ontologies in OWL and some factors to take into account when choosing your ontology editor. The next chapter builds on what we have learned so far and explains how to build a geographic ontology from the ground up, utilizing many of the OWL constructs introduced here.

NOTES

1. From this point, we indicate examples using Rabbit and Manchester Syntax shown side by side.
2. There is no concise Rabbit equivalent.
3. The "owl:Thing that" part of the sentence is often excluded as it is assumed as the default.
4. An aspect of the open world assumption means that the example given does not exclude the possibility that a semidetached house could not be attached to itself as it is a semidetached house. This is obviously silly. OWL does enable you to exclude this loophole by modifying the property "isAttachedTo." We show how this is done further in the book.
5. Assume these classes are mutually exclusive (disjoint).
6. In Manchester Syntax, individuals should really be URIs, but for brevity we are truncating them, so in the example Merea would really be of the form http://countries.data.world.org/Merea (or some such).
7. However, in some domains you may wish to model "part of" differently.
8. http://protege.stanford.edu
9. http://hermit-reasoner.com/.
10. http://owl.man.ac.uk/factplusplus/.
11. http://clarkparsia.com/pellet/.
12. http://www.comp.leeds.ac.uk/confluence/downloads.shtml
13. http://protegewiki.stanford.edu/wiki/ACE_View
14. http://protegewiki.stanford.edu/wiki/WebProtege
15. http://neon-toolkit.org/wiki/Main_Page
16. http://www.b2international.com/portal/snow-owl
17. http://knoodl.com/ui/home.html
18. http://www.topquadrant.com/products/TB_Composer.html
19. http://www.cognitum.eu/Products/FluentEditor/Default.aspx
20. http://www.racer-systems.com/.
21. See Appendix A for details of the OWL species and profiles.
22. http://code.google.com/p/elk-reasoner/.
23. http://lat.inf.tu-dresden.de/systems/cel/.
24. http://trowl.eu/.
25. http://www.altova.com/semanticworks.html

10 Building Geographic Ontologies

10.1 INTRODUCTION

In Chapters 6 and 7, we saw how Geographic Information (GI) can be represented as Linked Data using RDF (Resource Description Framework) and described at a basic level using RDFS (RDF Schema). This chapter builds on that baseline by showing how Linked Data can be more richly described using ontologies written in OWL (Web Ontology Language). It provides an overview of a method to build ontologies and then discusses the issues and techniques of ontology building with respect to GI. The techniques and problems dealt with are not exclusive to geography, but geography is distinguished by the degree and frequency to which certain issues occur. The techniques and approaches described here are therefore applicable in many other domains. The examples are expressed in the Rabbit controlled natural language, followed by the same information in OWL, using the Manchester syntax.

The OWL language itself was described in Chapter 9, and while this current chapter gives examples using many of the features described in Chapter 9, it does not present exhaustive examples of all the language features. This is because the emphasis in this chapter is to demonstrate how certain key characteristics of GI can be ontologically described, rather than attempting to present geographic examples of each OWL construct.

10.2 TYPES OF ONTOLOGY

Ontologies can be characterized in different ways, in overlapping categories, such as top level, domain, application, micro-ontologies, and link ontologies. While we have already mentioned link ontologies in association with link generation in Chapter 8, here we mainly deal with domain ontologies but also touch on top-level and micro-ontologies.

10.2.1 DOMAIN ONTOLOGIES

A domain ontology is one that describes the vocabulary and relationships that are associated with a particular domain of interest. It is application independent and will not contain terms specific to a particular application; rather, it will contain descriptions for the terms used in a general area of expertise. Organizations that supply general application-independent data, such as Merea Maps, are likely to produce domain ontologies. In the case of Merea Maps, they will be interested in developing a domain ontology that describes all the topographic things that are represented on their maps.

10.2.2 APPLICATION ONTOLOGIES

Application ontologies differ from domain ontologies as they include references to terms specific to a particular application or task. Domain ontologies will be used by application ontologies to draw on terms common to both the domain and the application; the application ontology will then add further terms that are related specifically to the application. We can imagine Merea Nature having an application to monitor the health of various habitats. It will construct the application ontology by taking elements of the topographic domain ontology constructed by Merea Maps, perhaps also a species definition from a domain ontology produced by an international wildlife organization, and its own domain ontology defining the habitats found on Merea. To this they will add terms specific to the application, possibly related to the monitoring of the habitats or events such as fire, drought, or flood that could affect the quality of habitats. The end result is something with a very specific purpose, less likely to be reusable by others than a domain ontology.

10.2.3 TOP-LEVEL ONTOLOGIES OR UPPER ONTOLOGIES

Top-level ontologies provide general vocabularies that can be utilized by domain and application ontologies. Top-level ontologies were some of the earliest to be developed and either tried to be encyclopedic in their coverage, such as OpenCyc,[1] or abstract in the extreme to generalize every entity into atomic concepts, such as SUMO[2] and DOLCE[3]; as a result, they are very large and unwieldy. The problem with these ontologies is that human knowledge is founded within context; these ontologies are not specialized and can become too philosophical. They use quite academic terminology like "endurant" and "perdurant" that are not in common usage and can be difficult to understand. It can become a challenge for the domain ontology author to continually try to fit the concepts in the domain ontology into the structure of the upper ontology. However, these ontologies do represent a considerable amount of thought and so will contain well-formed solutions to specific areas or knowledge modeling patterns, and these may well be worth reusing even if you do not want to use the entire ontology.

10.2.4 MICRO-ONTOLOGIES

Micro-ontologies are a more recent attempt to develop authoritative and reusable ontologies. These ontologies attempt to provide very specialized terms specific to well-defined domains and expertise. Examples are Dublin Core (International Organization for Standardization [ISO], 15836:2009) and the Spatial Relations ontology within GeoSPARQL (Perry and Herring, 2011). The first contains well-defined terms for describing metadata and the latter specific spatial relationships related to RCC8 (Region Connection Calculus 8). These are far more usable than top-level ontologies since they provide useful terms in small packages. Developing such micro-ontologies is generally good practice as it maximizes the chance of their reuse by others (the larger the ontology, the more likely you are to disagree with some terms within it even if you agree with others). It is also worth noting that ontologies such as Spatial Relations contain only property definitions, not class definitions; this is often a useful way to partition ontologies into modules.

10.3 METHODOLOGIES

There are a number of methodologies for developing ontologies. The benefit of using a formal methodology is that it provides structure to the process and helps to ensure best practice is carried out. We do not recommend a particular methodology but have identified METHONTOLOGY (Fernández-López, Gómez-Pérez, and Jursito, 1997), UPON (De Nicola, Missikoff, and Navigli, 2009), and Kanga (Mizen, Hart, and Dolbear, 2005; Dennaux et al., 2012) as options. They are all fairly similar in their approach, and UPON and Kanga have certainly borrowed from METHONTOLOGY. Both UPON and Kanga add competency questions (Noy and McGuinness, 2001) (questions that test that the ontology meets the purpose for which it is built); and UPON also adds use cases. Whatever methodology you adopt, we would recommend that at the very least the stages discussed next are used.

10.3.1 SCOPE AND PURPOSE

One of the most important and often missed stages is the very first: defining what the purpose of the ontology is and what its scope is. Without this, the construction of the ontology can lose direction and focus. If we take the Spatial Relations ontology as an example, we can say that the purpose is to provide a number of topological relationships that can be used by anyone needing to include such relationships in their applications and ontologies; we can say the scope is limited to RCC8 relations. The Spatial Relations ontology is a small micro-ontology, and the scope and purpose are quite easy to define. Defining the scope and purpose of domain ontologies can be more challenging—especially the scope, as it is all too easy to try to include more than is strictly necessary. Therefore, spending time up front thinking in detail about the scope and purpose is time well invested. Doing so can bring substantial savings in time later in the ontology authoring process, and indeed in helping to define an RDF vocabulary as identified in Chapter 7, and in the creation and linking of data-sets, covered in Chapter 8.

The scope and purpose will also help the authors to decide whether to use OWL or RDFS to describe the ontology. There may be circumstances when this is not necessary, most obviously if an organizational choice has mandated that all ontologies will be authored using a specific language. Otherwise, the scope and purpose will help to inform the choice of ontology language. The main criterion is how complex (descriptive) the ontology needs to be, and put simply, if it does not require the complexity of OWL, RDFS is a good choice.

10.3.2 USE CASES AND COMPETENCY QUESTIONS

Use cases and competency questions are particularly useful for application ontologies that have specific uses. Use cases are simply descriptions of the uses to which the ontology will be put; for application ontologies, these are typically the use cases of the application (e.g., monitoring the health of a habitat).

Competency questions are used to test whether the ontology meets its scope and purpose and effectively provide test cases for the ontology. There are two forms of question; the first tests for the completeness of the ontology; the second provides the basis to generate tests of the correctness of specific aspects of the ontology that can be translated into SPARQL or DL (Description Logic) queries. For the Spatial Relations ontology, a competency question to test completeness could be: Does the ontology contain all the RCC8 relationships and only these relations?

A test for correctness could be a test of the transitive nature of "contains" by specifying the expected behavior of the property. Specifically, the test could comprise the following:

Given:

```
A contains B.
B contains C.
D disjoint A.
D disjoint B.
D Disjoint C.
```

Then the test:

```
A contains ?x.
```

Should return B and C but not D.

They enable testing of the connectedness of the ontology. Ideally, the competency questions should be defined as early in the process as possible. It may be easier for the domain expert to couch the competency questions in terms of the application requirements; "Find all Parishes that contain Hospitals," for example.

10.3.3 Lexicon

The lexicon is constructed to build a list of the terms that need to be described within the ontology and that form the vocabulary. Typically, this is done by analyzing the content of documents related to the domain or application of interest and interviewing experts within the field. If there is doubt whether a term should be included in the lexicon, it is useful to hold the term up against the light of the scope and purpose.

10.3.4 Glossary

The glossary stage takes each term from the lexicon and develops an informal explanation expressed in natural language. This stage will also attempt to discriminate between core and secondary concepts. Core concepts are central to the scope of the ontology; secondary concepts are not strictly part of the domain or application but are required to help describe the core concepts. For example, in a topographic ontology Woodland will need to make a reference to Trees; without reference to Trees, it is difficult to describe Woodland, but equally Trees are not central to the topographic domain. So, while the topographic ontology will need to describe in detail the concept of Woodland, it will say little or nothing about the concept of a Tree, at most stating the difference between coniferous and broad-leaved trees. Thus, *in this case* Woodland is a core concept, and a Tree is a secondary concept. Merean Nature might take a different view as it will also need to be concerned about individual tree species, so both Woodland and Tree would be core concepts.

10.3.5 CONCEPTUALIZATION

The conceptualization stage is when the hard cognitive work happens and the natural language glossary is turned into an ontology of formally described classes and properties in an ontology language such as OWL. This stage will include testing the completeness and correctness of the ontology by applying the competency questions. One task that is becoming increasingly important as more ontologies are published is to check to see whether existing ontologies can be exploited to help build your ontology.

This description gives the impression that these stages are completely sequential; the reality is very different. Constructing the glossary can significantly overlap with the lexical analysis as the ontology author may choose to explain terms as they are discovered. The glossary stage will also be revisited during the conceptualization stage as new terms are discovered or some terms found to be unnecessary. In particular, this stage is likely to generate new terms for relationships between the various classes. So, expect the process to be very iterative.

In the next section, we work through these stages from the perspective of Merea Maps.

10.4 BUILDING THE TOPOGRAPHIC ONTOLOGY OF MEREA MAPS

10.4.1 SCOPE AND PURPOSE

Defining the scope and purpose can be surprisingly difficult, especially for domain ontologies. For Merea Maps to define the scope and purpose of its ontology, it first has to really understand what its purpose is and why it is building the ontology. This is not an easy question because in this case the ontology will not have a single specific application. Even where there is an initial application in mind, it may be difficult to specify scope and purpose simply because different people within the same organizations may have differing opinions regarding what the scope and purpose should be. In such cases, the process will help to arrive at consensus and will certainly highlight differences that might not otherwise be revealed until much later in the process. Defining the purpose should come first because if you do not know why you are doing something, the scope is irrelevant. So, why does Merea Maps want an ontology? Merea Maps sees itself as providing a referencing framework for other organizations and people, enabling them to use the features that Merea Maps provides for use within specific applications. For example, Merea Heritage can use the buildings recorded within Merea Maps' Linked Data and classify a subset of them as of historic interest, using the Uniform Resource Identifiers (URIs) provided by Merea Maps as the hook on which to link heritage-related data about those buildings. So, Merea Maps wants to provide a general topographic model of the Island of Merea and to publish this model as Linked Data (among other forms). The purpose of its ontology, then, is to provide a vocabulary that can be used to describe the model and explain what the data is. The topographic model and the ontology are then available for others. This means that the vocabulary should be of use not only to Merea Maps but also to those using its Linked Data. Merea Maps therefore defines the purpose of its ontology as follows:

> To provide a vocabulary that describes the Features and the relationships between them sufficient to enable the publication of topographic information as Linked Data and to be understandable by others so that they may query our data.

The scope then follows from the purpose. The scope refines the purpose by limiting the content to only those things necessary to fulfill the purpose. In terms of classes, this is relatively easy as the ontology needs to define core classes that relate to the Features that are surveyed by Merea Maps and the minimum set of secondary classes required to describe these core classes. Specifying the scope of the relationships is also done in terms of the minimal set of relationships required to describe the core classes. Merea Maps therefore defines the scope:

> To include all classes that describe the Features surveyed by Merea Maps and the minimal set of other classes and relationships necessary to describe these Features.

So, for example, if Merea Maps' ontology builders were considering whether to include the concept "University" in the ontology, then they could ask themselves, "Does it help to describe the topographic features in our Linked Data set?" to which the answer would be yes as Universities are shown on their maps. However, the term *University* itself is used ambiguously; it is used to refer to both the physical representation of the university (the buildings, land, roads, etc.) and the organization—the legal entity. Merea Maps' interest lies in the former not the latter. The description of the latter concept would be primarily outside the scope of a topographical ontology. While the University buildings are topographical features, the University as an organization is not a topographical feature. So, by specifying the purpose and scope of the ontology, Merea Maps' ontologists would be able to regard the University's physical manifestation as the primary concept and the University as an organization as a secondary concept and not add any details about the organizational features of the University.

10.4.2 Competency Questions

The level of detail of competency questions can vary. At their simplest, they can just be a set of general questions to make sure that the purpose of the ontology is met and that its scope is covered, such as, "Does the ontology include all topographic features in Merea Maps' data?" However, a more detailed approach can be taken that has some parallels to test-driven development, whereby more time is spent on crafting the tests, in this case the competency questions, than even on the ontology authoring itself to ensure that all the required reasoning is possible, and the expected outcomes are met. For example, a competency question might be "Find all the places with agriculture as a specified purpose that are adjacent to watercourses." To return the test results, the ontology needs to include concepts like Place, Agriculture, Purpose, and Watercourse, as well as relationships like "adjacent." But, we also need to make sure the links in the logic chain are included so that the question will return examples of Farms adjacent to Streams. That is, statements like "Every Farm is a kind of Place that has purpose Agriculture" and "Every Stream is a kind of Watercourse" will need to be included so that facts such as "Manor Farm is adjacent to Kingfisher Stream" will cause "Manor Farm" to be included in the results.

Where competency questions differ from traditional software engineering practices of test-driven development, however, is that we may well also be hoping for *unexpected* outcomes of a reasoner; particularly when integrating two ontologies, new information may be discovered serendipitously. It is therefore almost impossible to write questions to test for these unexpected outcomes. It is obvious that defining competency questions of this nature cannot be done early in the authoring process. It is important, though, that they are done as early as possible and are developed in line with the ontology, being built on and expanded as the development progresses.

10.4.3   BUILDING A LEXICON AND GLOSSARY

There are two principal sources that are used to build the lexicon and glossary: documentation and domain experts. The involvement of domain experts, either through direct involvement in the ontology authoring processes (something that we would strongly recommend) or indirectly through interviewing, can be especially enlightening. This is because documentary sources are often incomplete, out of date, and contradictory; documents that describe working practices or specifications may not define what is actually done (raising a separate question regarding which needs to be corrected): the guidance or practice. Both documents and domain experts will also provide descriptions that rely on assumed knowledge, so more detail may need to be teased out either through questioning of the domain expert or through reference to other material, such as dictionaries that provide a full definition.

Lexically, nouns typically identify possible classes of interest and verbs possible relationships (properties). Let us consider the following definition of a Duck Pond found in Merea Maps' *Guide for Field Surveyors*: "A duck pond is a pond that provides a habitat for ducks. Duck ponds may contain a duck house." Duck Pond, Pond, Habitat, Ducks, and Duck House are therefore all candidates for classes within the ontology and "is a," "provides," and "contain" are candidates for properties.

This description itself can probably be used to directly help construct the glossary entry for Duck Pond as we can easily determine that Duck Pond is a core concept (since the Guide for Field surveyors defines the things that surveyors need to record). Similarly, Pond is also a core concept, and we would expect to find this defined elsewhere in the guide. Habitat and Duck, however, are not core concepts, so we would not expect Merea Maps to provide a detailed description, and the glossary will merely identify them as secondary objects. Duck House is more challenging. It is clearly a physical object but does not appear elsewhere in the surveyor's guide. It can therefore be assumed that it is not considered significant enough to be recorded and so will be specified as a secondary concept. Sometimes, it is worth double-checking this type of thing as often documentation contains omissions or does not reflect current practice.

Turning our consideration to the candidate properties, it is quickly realized that "is a" corresponds to the subsumption property that comes by default with RDFS as `rdfs:subclassOf`.[4] The "provides" appears to be a good solid candidate for a property, but "contains" is not, at least not on the evidence provided by this single example; we are only concerned with describing the essentials of a duck pond that are applicable to all duck ponds, not optional attributes that may only apply to some duck ponds.

One area that requires careful thought is where something is known by different names, especially if this is related to a geographic distribution. Multiple names for the same thing are not unusual; on Merea, Pavement and Sidewalk are used more or less interchangeably throughout the island. Therefore, although in documentation or from our domain experts we may identify the two nouns, we have only one class with two synonyms. Modeling them as two separate classes and then making them equivalent would be incorrect as the difference is strictly lexical.[5] Here, the glossary will need to define the terms as synonyms, and it is represented within the ontology as one class with two labels:

```
Pavement is a concept that has a synonym Sidewalk.      Class: Pavement
SubClassOf (
Annotation("Sidewalk")).
```

Harder to resolve is where the thing is fundamentally the same but the geographic distribution is distinct. Consider this: In North Merea, small streams are consistently referred to as brooks; in South Merea, they are consistently known as becks. So, do we have two different classes or one class with two synonyms? It all very much depends on how important it is to discriminate between the two; if it is not that important, then it will be sufficient to treat the terms as synonyms; if the geographic difference is important, then two classes are required. Merea Maps needs to ask the question: "If a user asks for becks, will the user be surprised if brooks are returned as well?"

```
Every Brook is a kind of Small Stream.
Every Brook is found in North Merea.
Every Beck is a kind of Small Stream.
Every Beck is found in South Merea.                     Class: Brook
SubClassOf: SmallStream that isFoundIn value northMerea
Class:Beck
SubClassOf: SmallStream that isFoundIn value southMerea.
```

10.4.4 DESCRIBING CLASSES

Once the glossary is substantially defined it is possible to start to build the actual ontology by formally describing the classes and properties (relationships). The very first question that Merea Maps asks is, "Do ontologies already exist that either do what we need or have components that we can use?" It finds that there are some ontologies, largely micro-ontologies, that describe useful properties, and Merea Maps decides to use these rather than reinvent the wheel.

These are the following:

Ontology	Description	Prefix
OGC (Open Geospatial Consortium) Geometry	Contains definitions for the basic OGC geometry primitives corresponding to points, lines, and polygons	OGC:[6]
RCC8 Topology	Contains definitions of the RCC8 relations	RCC8:
Mereology	Contains basic part of and whole relations	GeoParts:
Geographic Names	Defines a class for geographic names and associated relations	GeoName:
Network Topology	Contains basic network topology classed (link and node) along with standard network relations such as connects to	NetTopo:

Having selected some ontologies to reuse, Merea Maps now embarks on the main task of constructing the ontology; the first part of this activity is usually the hardest and takes up a disproportionate amount of time, especially for domain ontologies. This is because the first phase is about establishing high-level classes and design patterns or templates that can be applied to many of the classes. It is particularly true of domain ontologies as these often contain many classes that can be grouped and expressed using particular patterns.

10.4.4.1 Top-Level Classes

Establishing the top-level classes in an ontology can be deceptively easy, but if done properly quite hard. This is certainly true for Merea Maps. Merea Maps already has a Feature Type Catalogue that it uses to define the content of its maps, and this catalogue is arranged as a hierarchy starting at the top with "Map Feature" and going all the way down to the individual surveyable objects such as buildings, roads, rivers, factories, schools, and so on. The deceptively easy bit is simply to use this hierarchy to establish the basic structure of the ontology. But rather than just doing this, Merea Maps pauses and asks whether a hierarchy defined thirty years previously to meet the needs of surveyors is suitable to be used as is by an organization wishing to publish data on the Semantic Web. The answer it comes up with is that while much of the lower levels of the hierarchy can be reused, the top level requires serious rethinking. The reason for this is twofold. First, while the lower levels of the hierarchy refer to fairly concrete concepts such as Building, the levels above become progressively more abstract and thus open to philosophical dispute. In the case of Merea Maps, the hierarchy divides early on into two branches, one for natural features, the other for artificial features. This split may seem fairly logical, but in truth it can be very difficult to determine whether something is artificial. In Merea, there are very few things that have not been affected or altered by the action of people. Even seemingly complete natural features such as woods and rivers have been altered over the passage of time by the islanders; rivers have been straightened and rerouted, and all the woods have been managed, Merea of the twenty-first century has no wild wood. This leads to the second factor: In an ontology, everything must add value; the upper layers must provide useful components that add descriptively to the lower levels or that are useful to end users querying and interacting with the data. What benefit does differentiating between natural and artificial features bring? The answer that Merea Maps has arrived at is that it adds very little benefit; the classification is not only difficult to unambiguously specify but also adds nothing that can be reused in a concrete sense by the lower levels. As a result, they set about redesigning the top level of their hierarchy such that it is now suitable to be used in an ontology. In fact, the process of constructing this top level is not one that can usually be done simply in a top-down manner. Some things are obvious—everything of primary interest to Merea Maps has a location, so we can expect a high-level class. In the Merea Maps case, Topographic Feature (or Feature for short) will specify a location property. Merea Maps states that Features have a Footprint:

```
Every Feature has a Footprint.    Class: Feature
                                  SubClassOf: hasFootprint some
                                  Footprint
```

The Footprint class in turn comprises one or more OGC geometries that describe areas, points, and lines related to Earth's surface.

```
Every Footprint has geometry      Class: Footprint
one or more of OGC Point, OGC     SubClassOf: hasGeometry some
Line, or OGC Polygon.             (OGCPoint or OGCLine or OGCPolygon)
```

This means that any number of geometries can be associated with a Feature. The reason why more than one geometry may be necessary is that, in the case of Merea Maps, it wants to be able to specify the geometry at different scales, including just a point reference for very small scales.

Next, Merea Maps tries to see if there are other top-level classes that are immediate subclasses of Feature. To do this, the nature of the Features that they survey are examined to see if there are any obvious groups and to see if these groups are useful to the end users. They come up with a number of potential candidates: Administrative Areas, Settlements, Landforms, Structures, and Places as shown in Figure 10.1.

Most of these classes are obvious; Place is less obvious and, as we shall see, is also not the best choice. Although there are many different definitions for Place, Merea Maps decides to use it in a very specific way: A place is somewhere (a Feature) where there is a designed purpose (intent) for something to happen there. They then use it to cover things such as hospitals, golf courses, and other complex areas where some specific activity is intended to take place. At this point, one of the main things that these classes do is to split the ontology into manageable chunks where it is reasonably easy to assign subclasses in an unambiguous way. In fact, if Merea Maps believes that it can unambiguously assign particular types of things such as farms or towns to one and only one of these high-level classes, then it can explicitly make these classes disjoint, for example:

```
Administrative Area and Settlement are    Class: AdministrativeArea
 mutually exclusive.                       DisjointWith: Settlement
```

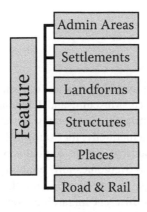

FIGURE 10.1   Initial top level of the ontology.

But Merea Maps needs to be very sure that this is always true. Certainly, in the case of Administrative areas and Settlements it is not only true, but also making them mutually exclusive can help to detect confusions that may exist between them. (A very common mistake is to believe that the settlement and the administrative area that is responsible for the settlement are the same thing.) However, overuse of disjoints can slow a reasoner, so caution should be exercised before use.

As we shall see, these top-level classes also provide common properties for their subclasses. The development of the top-level classes is likely to be quite iterative, especially in the early stages as design patterns are established.

Before we start to describe the next phase of developing design patterns and establishing lower-level classes, we stop for a moment to point out that an ontology is rarely a simple hierarchy; the majority of it is a network. We have already encountered this with the Footprint class. Footprint does not sit within the hierarchy of Features but is nevertheless an important component of the ontology. Ontologies normally comprise a number of interleaved hierarchies that form a network; establishing these hierarchies and the interrelatedness of them is an important part of the ontology authoring process. One should not feel limited to one single hierarchy (a taxonomy).

As a general rule, it is good design practice to have each hierarchy as shallow as possible. The more high-level classes there are, the more difficult it may be for a third party to reuse the ontology. This is again because higher-level classes tend to be more abstract and therefore more open to disagreement. It is also important to note that hierarchical relationships are just that; they are strict trees and so can only express one type of relationship: subsumption, "is a kind of." The problem is that if subsumption is the only tool you have, and in traditional hierarchical classification systems this is all you have, it is very easy to mix categories from different hierarchies. For example, it is very common to see the following type of "hierarchy":

```
Topographic Feature
        Business & Industry
                Factory
                Office
            ...
```

```
Tourism & Entertainment
        Museum
        Cinema
        ...
```

The issue is that Business & Industry and Tourism & Entertainment are functions, whereas Factory, Office, Museum, and Cinema are things—two different types of categorization are mixed. This creates the classic problem of not being sure where something fits in a hierarchy; for example, Cinema could also be reasonably put under Business & Industry. An ontological approach allows us to recognize these different hierarchies and overcomes it by enabling us to create different types of association between things; "is a kind of" is not the only tool in the box. In the example, we can create two different hierarchies, one of things and the other of functions, and relate them through some properties:

```
Topographic Feature
        Factory
        Office
        Museum
        Cinema
        ...
Function
        Business
        Industry
        Tourism
        Entertainment
        ...
Cinema has function Business.
Cinema has function Entertainment.
```

Because Merea Maps is taking an ontological approach, it can avoid putting lots of different meanings into a single hierarchy.

10.4.4.2   Developing the Detail: Places

Merea Maps begins the process of describing the detail of their ontology by starting with Places. To remind ourselves, Merea Maps describes a Place as follows: "A place is somewhere (a Feature) where there is a designed purpose (intent) for something to happen." Going into more detail, the ontologists discover that Places have been created by Merea Maps for a particular primary purpose and are things such as farms, hospitals, factories, playing fields, and so on.

The first thing they need to do is assign subclasses to this class, and they are able to do this from their Glossary, selecting those Features that seem to have a designed purpose, such as factories, farms, hospitals, and golf courses. They then take each of these subclasses in turn and build a description for it. So, if they start with Farm:

```
Every Farm is a kind of Place.      Class:Farm
                                    SubClassOf: Place
```

and by saying that a Farm is a Place, the Farm class inherits location (Footprint) indirectly through Place, which in turn inherits this axiom from the Feature class.

This is the easy bit. But, what else can Merea Maps say about a Farm? At the very least, we know that a Farm must physically have at least one building and a number of fields.

```
Every Farm has part a Building.      Class: Farm
Every Farm has part a Field.         SubClassOf: hasPart some
                                       Building, hasPart some Field
```

Note that we have chosen to use the mereological relationship "has part," not the topological relationship "contains," so we are saying that a Farm has a Building as a part, rather than simple containment. Beyond this, there is very little that is physically common to all farms—even things that might be common to the vast majority of farms may not be true of all of them. For example, we might expect a farm to have lanes and paths, but it would be possible to have a farm that comprised a single building set in a field. The farm would not exactly be very big and only marginally a farm but would be a farm nonetheless. This emphasizes an important point: Ontologies provide descriptions that apply to all instances of a class of thing, not typical descriptions that apply to most instances. So, we only say what every farm must have—the necessary conditions for being a farm—and we do not include things that most farms have, but some do not, or things that farms could have. The descriptions of classes are therefore minimalistic. This is a common cause for misunderstanding and frustration when building an ontology. The first thing to remember is that not saying that a class can have a particular property does not mean that an individual of that class cannot have that property; it just means it is not common to all individuals. So, just because Merea Maps' class of Farm does not include a reference to lanes does not mean that a particular farm cannot have a lane. If we want to exclude something explicitly, we have to say so. This is the open world assumption rearing its head again. It is reasonable to suppose that no farm will have a steelworks as an integral part of the farm, so we could say if we so wished that

```
No Farm has part a Steel Works.[7]
```

However, in reality, to avoid having to mention all the impossible objects that are not a part of a farm, we refer to the scope and purpose of the ontology and ask ourselves why we need to include the fact that farms do not have steelworks. We will probably find that there is no good reason to state this fact even if it is true.

Farms do not exist without a reason; they have a purpose (intended use), and Merea Maps uses this as another factor to describe the farm class.

```
Every Farm is intended for      Class: Farm
  Agricultural Production.        SubClassOf: isIntendedFor some
                                    AgriculturalProduction
```

In fact, Merea Maps has now reached a point where it has a minimalistic description of a farm.

```
Every Farm is a kind of Place.          Class: Farm
Every Farm has part a Building.         SubClassOf: Place that hasPart
Every Farm has part a Field.              some Building and hasPart some
Every Farm is intended for                Field and isIntendedFor some
  Agricultural Production.                 AgriculturalProduction
```

A farm is something that has a specific location, comprises at least a building and a field, and is intended for Agricultural Production. As you can see from the Manchester OWL syntax, we take the union (using either "and" or "that") of these sentences describing a Farm, and the class Farm becomes a subclass of those Places that also have a Building part and a Field part and are intended for Agricultural Production.

At this point, it is worth noting that Merea Maps has reused the class Place described previously in the ontology and reused the "has part" property from the Mereology ontology. It has also introduced the new classes of Building, Field, and Agricultural Production along with the properties "has a part" and "is intended for"; at this stage, they have not described them further even though they are either core classes or properties.

Merea Maps next choose to describe a school. Again, they can say that it is a Place and also that it must comprise at least one building. It also has a well-defined purpose: that of providing education. So, Merea Maps can therefore provide the following minimal description of a school:

```
Every School is a kind of Place.        Class: School
Every School has part a Building.       SubClassOf: Place that hasPart
Every School is intended for              some Building and isIntendedFor
  Education.                               some Education
```

10.4.4.3 Developing Patterns

From this, it can be seen that a pattern is beginning to emerge: that of a combination of location, physical construction, and purpose. Indeed, we find that this pattern can be reused for many different classes of things—factories, hospitals, universities, airports, and so on. All of these will have a specific location, buildings, and purpose. A common factor of many ontologies is that class descriptions conform to a relatively small number of patterns. Such patterns are more normally discovered rather than designed from the outset, but an important thing to remember is that they will not be discovered unless you look for them.

In the examples given so far, all have buildings, but there are other things where buildings are not mandatory—a golf course, for example. Many golf courses will have buildings (a clubhouse, maintenance buildings, and so on), but none is mandatory. So, we can provide the following description:

```
Every Golf Course is a kind of          Class: GolfCourse
  Place.                                 SubClassOf: Place that hasPart
Every Golf Course has part Golf           some GolfLinks and isIntendedFor
  Links.                                   some Leisure
Every Golf Course is intended for
  Leisure.
```

In general terms, the pattern itself can be expressed as

```
Every <X> is a kind of Place.      Class: X
Every <X> has part a Feature.      SubClassOf: Place that hasPart
Every <X> is intended for a          some Feature and isIntendedFor
  Purpose.                           some Purpose
```

Merea Maps decides to add these general properties to the description of a Place. It decides that Purpose can represent a new hierarchy (and potentially a new micro-ontology). So, we can now define Education, Leisure, and Agricultural Production as Purposes:

```
Education is a kind of Purpose.      Class:Education
Leisure is a kind of Purpose.        SubClassOf: Purpose
Agricultural Production is a kind    Class:Leisure
  of Purpose.                        SubClassOf: Purpose
                                     Class:AgriculturalProduction
                                     SubClassOf: Purpose
```

A Place can be described as:

```
Every Place is a kind of Feature.    Class:Place
Every Place has part a Feature.      SubClassOf: Feature that hasPart
Every Place is intended for a          some Feature and isIntendedFor
  Purpose.                           some Purpose
```

And, a School can now be described as:

```
Every School is a kind of Place.     Class:School
Every School has part a Building.    SubClassOf: Place that hasPart
Every School is intended for           some Building and isIntendedFor
  Education.                           some Education
```

However, Merea Maps is not totally happy with this because what it does not do is relate the purpose of the place to the essential components of the place. Specifically, the description of a school says a school must have at least one building but does not relate the purpose of the building to education. So, what is required is for the building itself to specify its purpose. We would modify the description of the school as follows:

```
Every School is a kind of Place.     Class:School
Every School has part a Building     SubClassOf: Place that hasPart
  that is intended for Education.      some (Building that
Every School is intended for           isIntendedFor some Education)8
  Education.
```

We still have to explicitly say that the School has the intended use of Education because although we have now also said this about a building within the School, this statement would not imply the School as a whole had that as its purpose.

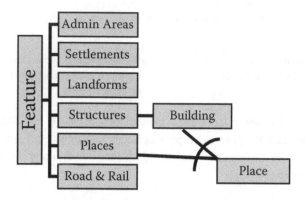

FIGURE 10.2   The revised top level of the ontology.

Once we have done this, we can therefore better describe Place as follows:

```
Every Place is a kind of Feature.      Class:Place
Every Place has part a Feature that    SubClassOf: Feature, (hasPart
is intended for a Purpose.              some Feature that
Every Place is intended for a Purpose.  isIntendedFor some Purpose),
                                        isIntendedFor some Purpose
```

On the face of it, this seems perfectly logical, but all is not what it may seem. The problem is that there is nothing that states that the purpose of the Feature that is a part of the Place has to be the same as the purpose of the Place itself. This problem can be resolved, as we discussed in Section 10.2.14.4, but even this is not an ideal solution. For now, Merea Maps decides to settle with this solution because although it is imperfect, it decides that it is still a useful pattern and does enable general queries to be constructed.

A further issue is that a Place can of course be just a single Feature; a school might comprise a single building and nothing else. The description of a school used previously and the general pattern will deal with this, but in these cases it does introduce a level of complexity that is strictly unnecessary for such Features; that is the School is a Place that contains a single building, as opposed to just a building. One way around this is to redefine Place by making it either a Site or a Building, where a Site has the old definition of Place (Figure 10.2).

```
Every Site is a kind of Feature.       Class: Site
Every Site has part a Feature that     SubClassOf: Feature, hasPart
is intended for a Purpose.              some Feature that is
Every Site is intended for a Purpose.   IntendedFor some Purpose
Every Place is a kind of Site or       Class:Place
Building.                               SubClassOf: Site or Building,
Every Place is intended for a Purpose.  isIntendedFor some Purpose
```

Here, Site now represents any collection of Features that are related by some Purpose, and a Site may have no buildings but does not exclude the possibility of buildings being present. A Place is now either a kind of Site or a kind of Building. Place has to repeat the intended Purpose as this is necessary to ensure that if a Place is just a kind of Building, then it also has a Purpose associated with it.

10.4.4.4 Use and Purpose

This pattern for Place cannot be applied to all Features because while all Features can be described in terms of the required parts, not everything has a purpose. What is the purpose of a hill, river, or cave? All of these things can have uses but they do not have a purpose (or at least not as a rule; it is possible that a specific hill, river, or cave could be an artifice and have an intended purpose, but these are atypical as most are natural landscape features). Here, the idea of *use* has been introduced, and it can also be seen that artificial features can have uses: A school can be used not only for educational purposes but also as a voting station, an emergency center, and so on. Whether the Feature has a purpose or not, what is clear is that uses can only be applied to particular and specific instances of things. We can say that a particular river such as the River Adder is used for canoeing, but we cannot say all rivers are used for canoeing. Similarly, the Merean government has allocated only certain schools to act as emergency centers, not all the Merean schools. What this tells Merea Maps is that it needs a new property "has use," but unlike the property "is intended for," "has use" will not be used in statements within the Topographic Ontology that it is developing, although it will be described as a property within the ontology. This is because "has use" is only applied to individuals of classes, not to classes themselves. Merea Maps also realizes that there is a relationship between "has use" and "is intended for"; the latter is a subproperty of the former, and indeed Purpose should really be renamed Use. To understand why it is advantageous to form this relationship, consider a football stadium. We can say that all football stadia have the intended purpose of leisure activity and more specifically playing football; this also tells us that "playing football" is a subclass of "leisure activity." Now, a specific football ground, Medina Academicals Football Stadium, is also used as a music venue, but this is not its main purpose; performing music is not what makes a football stadium a football stadium—playing football is. Merea Maps can therefore describe the Medina Academicals Football Stadium as follows:

Use is a Concept.	Class: Use
"has use" is a relationship.	ObjectProperty: hasUse
"is intended for" is a special	ObjectProperty: isIntendedFor
type of "has use".	SubObjectPropertyOf: hasUse
Football Pitch is a kind of	Class: FootballPitch
Feature.	SubClassOf: Feature
Leisure Activity is a kind of Use.	Class: LeisureActivity
Playing Football is a kind of	SubClassOf: Use
Leisure Activity.	Class: PlayingFootball
Performing Music is a kind of	SubClassOf: LeisureActivity
Leisure Activity.	Class: Performing Music
Every Football Stadium is a kind	SubClassOf: LeisureActivity
of Place.	Class: FootballStadium

```
Every Football Stadium has part a        SubClassOf: Place, hasPart some
  Football Pitch.                          FootballPitch, isIntendedFor
Every Football Stadium is                  some Playing Football
  intended for Playing Football.         Individual: http://data.mereamaps.
Medina Academicals Football               gov.me/medina_academicals
  Stadium is a Football Stadium.         Types: FootballStadium
Median Academicals Football              Facts: hasUse some PerformingMusic
  Stadium has use Performing Music.
```

Now because "is intended for" is a subproperty of "has use," we can ask both what the intended purpose of the football stadium is ("playing football") and what its uses are ("playing football" and "performing music").

Merea Maps also does one other significant thing: It realizes that it would be sensible to create a new micro-ontology "Uses" that will contain the usage hierarchy that it has started to develop along with the two properties "is intended for" and "has use." By doing so, it enables others to use this ontology on its own, if they so wish, and ensures that the Topographic Ontology itself does not become too bloated. This whole process has also shown that what we are beginning to see is the construction of a network through the interaction between separate hierarchies: so far a hierarchy describing the relationships between Features and a hierarchy describing Usages. The Feature hierarchy is also used twice in the construction of the network describing places.

10.4.4.5 Other Ontology Design Patterns

An ontology design pattern is a "reusable successful solution to a recurrent modelling problem," and so far in this section, we have described those that have arisen in Merea Maps' ontology. However, it may well be that you encounter the need for other ontology design patterns as you proceed to construct your ontology. While it is beyond the scope of this book to describe all the patterns that have been suggested in the literature, as they have arisen in attempts to solve modeling problems in many different domains, there is a useful Web site listing such solutions: http://ontologydesignpatterns.org. These include patterns like "punning," "*n*-ary relations," "value partitions," and part-whole relations.

Punning addresses the problem of wanting to refer to something as both a class and an instance in the same ontology, depending on context. This is known as "metamodeling," and while it is straightforward to do in RDFS or OWL Full, it can only be achieved in a DL-compatible fashion in OWL 2 by declaring the thing as a class and then reusing it as the subject of a statement involving an object or data property. The OWL reasoner then treats the class and individual views of the object as different things, although they share the same URI.

Rather than the usual binary relation that links one individual to another individual or value, an ***n*-ary relation** links an individual to several other individuals or values. This allows us to include more information about the linking property, for example, our certainty about it, its relevance, or context; or, we can model the links between the subject, direct object, and indirect object of a statement.

There are four main use cases for *n*-ary relations, which can be represented by different modeling patterns (Noy and Rector, 2006):

1. Adding an extra argument to a binary relationship. For example, Ash Fleet Farm is worth 2 million Merean Schrapnels with a high probability. This can be modeled by turning the relationship "worth with probability" into an individual of an anonymous class, WorthRelation1. The WorthRelation1 individual then has its own properties, *worth_amount* and *with_probability*. Both these properties are defined as functional properties, so that Worth Relation1 has exactly one worth amount and one probability (Figure 10.3).

2. When two binary properties always occur together and ought to be represented as an *n*-ary relation. For example, Water Pollution in the Isis is low but increasing. This can be modeled in the same way as the previous example, so the anonymous class could have an instance Measurement1, with two functional properties has_absolute_measurement and has_trend (Figure 10.4).

3. When the relationship links several objects. For example, Ash Fleet Farm was sold to Farmer Giles for 250 Merean Schrapnels in 1976. In this case, there is no clear subject for the *n*-ary relation, so the pattern looks like that in Figure 10.5.

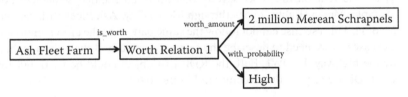

FIGURE 10.3   *n*-ary relations: Ash Fleet Farm is worth 2 million with a high probability.

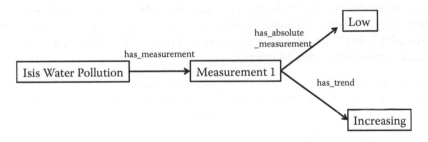

FIGURE 10.4   *n*-ary relations: Isis Water Pollution is low but increasing.

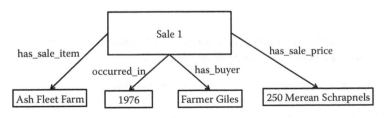

FIGURE 10.5   *n*-ary relations: Ash Fleet Farm was sold to Farmer Giles for 250 Merean Schrapnels in 1976.

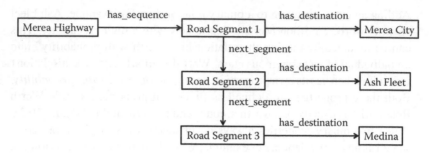

FIGURE 10.6 *n*-ary relations: Merea Highway's road segment sequences.

These three examples all create individuals belonging to anonymous classes, and we have not given meaningful names to either the individuals or those classes because they are never used on their own, but only as part of the pattern to model the *n*-ary relation. Therefore, there is no reason to invent a distinguishing name.

4. When the relationship links several objects in an implied sequence. For example, Merea Highway goes through Merea City, Ash Fleet Village, and Medina. This use case cannot follow the same pattern as the aforementioned use cases as we need to describe the sequential nature of the cities reached by the highway. In OWL Full, the RDF List object could be used, but for OWL DL we require the structure in Figure 10.6.

10.4.4.5.1 Value Partitions and Value Sets

The value partitions and value sets of patterns are useful when you want to model adjectives or other modifiers, for example, an "expensive beer" or a "three-star pub," where there are only a limited number of choices (e.g., "cheap," "moderate," and "expensive"). In OWL, there are a few ways to approach this, either as a partition of classes (known as a "value partition"):

```
Beer is a Concept.                Class: Beer
Porter, Bitter, Stout, Lager, and    DisjointUnionOf: Porter, Bitter,
Ale are Concepts.                    Stout, Lager, Ale
Every Beer is exactly one of
 Porter or Bitter or Stout or
 Lager or Ale.
Porter, Bitter, Stout, Lager, and
 Ale are mutually exclusive.
```

or at the individual level, known as "a value set", by enumerating the individuals or listing all the values using datatype properties:

```
PubRating is a Concept.           Class: PubRating
Every PubRating is exactly one of    ObjectOneOf: onestar, twostar,
 onestar or twostar or threestar.    threestar
onestar, twostar and threestar       DifferentIndividuals: onestar,
 are different things.               twostar, threestar
```

10.4.4.5.2   Part-Whole Relations

There are a number of design issues that should be taken into account when modeling mereology (part-whole relations), many of which might be encountered in GI ontologies. First, we should note that the two basic relationships hasPart and isPartOf are inverses of each other. On an individual level, with individuals I1 and I2, if we state that "I1 hasPart I2" and that isPartOf is the inverse of hasPart, then we know that "I2 isPartOf I1." However, this reciprocal inference does not hold at the class level. For classes A and B, the statement "Every A hasPart some B" and isPartOf is the inverse of hasPart does not allow us to assume that "Every B isPartOf some A"; rather it must be stated explicitly, which adds a lot to the reasoner's workload. Usually, it is best to decide to use either isPartOf or hasPart, depending on how the ontology will be used and what the competency questions are.

In traditional mereology, the partOf relation is transitive, reflexive, and antisymmetric. The antisymmetric property means that "Nothing is a part of its parts" (recall that this is slightly different from asymmetry, which states that if A is related to B, then B cannot be related to A). However, antisymmetry cannot be represented in OWL 2. In mereology, since everything is a part of itself (reflexive), we need to add in the term "hasProperPart" to deal with every part except the whole. In modeling, one also needs to be careful that isPartOf is used only when it is really meant. Some relationships that can be confused with mereological ones include

- Subclass ("is a kind of")
  For those used to modeling vocabularies, the superclass/subclass hierarchy is the only relationship on offer and is hence overused when in fact a different, more meaningful relationship, such as "is a part of," is required.
- Containment ("is inside of", "is contained in")
  An object may be inside something, but this does not mean it is part of it. For example, the water contained in a glass is not part of the glass.
- Membership ("is a member of")
  Being part of a group actually refers to being a member of that group. While "is part of" is transitive, membership is not. For example, while my arm is part of me, it is not part of the committee I sit on—because the latter relationship is one of membership.
- Constituents ("made of," "is a constituent of")
  While a statue may be made of stone, the stone is not part of the statue with the same meaning as an arm is part of the statue.
- Connections and branches ("is connected to")
  We may want to distinguish between things that are connected to a system, say a railway network, and things that are part of the network itself. So, in an integrated transport network in a city, for example, a bus route will be connected to the railway network, although it is not part of the network itself. More subtly, a lamp connected to the electricity system is not part of the electricity system.

All these alternatives to "part of" offer ways of adding nuanced meaning to an ontology.

10.4.4.6 Breaking the Pattern Rules

Before concluding this discussion on patterns, we sometimes come across other things where a pattern applies but we do not think they belong to the class that defines the pattern. Merea Maps encounters this when describing Landform classes. It starts to describe an Orchard:

```
Every Orchard is a kind of            Class: Orchard
  Enclosed Land.⁹                     SubClassOf: EnclosedLand, hasPart
Every Orchard has part Trees.           some Tree, isIntendedFor some
Every Orchard is intended for           AgriculturalProduction
  Agricultural Production.
```

and realizes that this looks very like the pattern created for Place. So, is an Orchard really a Place and not a Landform, is it a Place *and* a Landform, or does Merea Maps still consider it to be only a Landform? Once again, there is no absolute answer, merely different solutions that may apply in different contexts. Strictly, Merea Maps could argue that as an Orchard is somewhere where something was intended to happen—in this case growing fruit for agricultural purposes—then an Orchard must be a Place. However, Merea Maps is also aware that the majority of its users would naturally classify an Orchard as a Landform, not a Place; it also does not like the idea of making it a subclass of both Place and Landform and so settles on describing it as it has. The reader may disagree with this decision and choose a different solution. This just highlights that all ontologies are imperfect compromises, and that there are often situations for which there are no best solutions. It also demonstrates that top-level classes such as Place and Landform can be really quite weak, with different people having different ideas over membership.

10.4.5 PROPERTIES

We have so far largely discussed the construction of the ontology from a concept- or class-oriented view. This is quite natural as people have a tendency to focus on things and then think about how these things relate to each other; people do not tend to think of properties or relationships and then identify the things that can be associated using these properties. However, the development of properties is just as important as the classes. Using the example of social networks or transport networks, we can see that the links between the people or transport hubs are as informative as the descriptions of the people or hubs themselves. In terms of the pragmatics of ontology authoring, they are developed hand in hand with the classes they help to describe. Although Merea Maps has begun creating properties such as "is intended for," "has use," and "has part," it has not seriously thought too much beyond the subproperty relationship between "is intended for" and "has use." We have seen in Chapter 9 on OWL that, compared to RDFS, OWL introduces additional richness to the way we can describe a property.[10] As Merea Maps proceeds through its ontology, it becomes more aware of how it can improve the ontology and indeed the underlying data by developing more sophisticated properties.

10.4.5.1   Symmetric Properties

In Chapter 6, it was shown how Merea Maps could represent a road network using statements of the form:

```
Medina is connected to Medina Road.    Individual: Medina
                                       Facts: isConnectedTo MedinaRoad
```

This just says that Medina is connected to Medina Road, not that Medina Road is connected to Medina. The following additional statement was required:

```
Medina Road is connected to Medina.    Individual: Medina Road
                                       Facts: isConnectedTo Medina
```

This is because in RDFS there is no way to say a property works both ways around, and as a result we have to produce twice as many assertions as we would like. From Chapter 9, we have seen that OWL provides a way to make the "is connected to" property work in both directions. Merea Maps achieves this by stating that the property is symmetric:

```
The relationship "is connected to"     ObjectProperty: isConnectTo
 is symmetric.                         Characteristics: symmetric
```

Now, it is simply possible to write one or the other of the statements about the connectivity between Medina and Medina Road, and the other is automatically inferred.

10.4.5.2   Inverse Properties

Merea Maps then realizes it not only has a similar issue with "has part," but also that it is not quite the same. What it would like to be able to do is only say either X is a part of Y, or that Y has a part X, and not have to say both statements all the time. However, this is not identical to the "is connected to" property, and making "has part" symmetric will not work. This is because the "has part" property has a distinctly different meaning from "is a part of," whereas "connected to" worked in both directions. However, Merea Maps can see that OWL still allows it to do what it wants by making the property an inverse property rather than a symmetric one as described in Chapter 9. This is demonstrated in the specific example that follows. Merea Maps knows that Lower Field is part of Ash Fleet Farm; therefore, Ash Fleet Farm has a part that is Lower Field. Rather than having to explicitly state both facts, Merea Maps now just states:

```
The relationship "part of" is the      ObjectProperty: partOf
 inverse of "has part".               InverseOf: hasPart
Lower Field is part of Ash Fleet Farm. Individual: LowerField
                                       Facts: partOf AshFleetFarm
```

Then, the following statement can be inferred:

```
Ash Fleet Farm has part Lower Field.   Individual: AshFleetFarm
                                       hasPart LowerField
```

10.4.5.3 Transitive Properties

Having defined "has part" to be an inverse property, Merea Maps then realizes that it can further enrich "has part" by making it transitive, although at first little realizing the implications. A field that is part of Ash Fleet Farm, which in turn is part of Ash Fleet Farm Estate, is obviously also part of Ash Fleet Farm Estate. Merea Maps supports this inference by making its "has part" property transitive:

```
"has part" is a relationship.     ObjectProperty: hasPart
The relationship "has part" is    Characteristics: Transitive
  transitive.
```

However, in doing so any query that asks "What are the parts of Ash Fleet Farm Estate?" will get back everything, not just its immediate components: the Farm and Ash Fleet House. This is a bit like asking what makes up a car and getting every component down to the last nut and bolt back, rather than just the major elements, such as the chassis, wheels, engine, body, and so on. There are occasions when getting back a complete component list is what is desired, but more often than not this is not what is required. To some extent, this can be mitigated by specifying particular class types in a SPARQL query so that the query does not return things that belong to class types referring to low-level parts, but this could get a bit messy and may not always be possible. The transitive restriction is a very blunt weapon. Merea Maps wants both the ability to return all components and just those that are the immediate subcomponents of a Feature. They are able to do this by devising a subproperty of "has part" that is not transitive:

```
The relationship "has direct       ObjectProperty: hasDirectPart
  part" is a special type of the   SubPropertyOf: hasPart
  relationship "has part".
```

The trick here is that only domain and range restrictions are inherited; hence, "direct has part" is not transitive, even though its parent property is.[11] Merea Maps then uses this subproperty rather than "has part" as follows:

```
Ash Fleet Farm Estate has direct   Individual: AshFleetFarmEstate
  part Ash Fleet Farm.             Facts: hasDirectPart AshFleetFarm
Ash Fleet Farm has direct part     Individual: AshFleetFarm
  Lower Field.                     Facts: hasDirectPart LowerField
Ash Fleet Farm has direct part     Individual: AshFleetFarm
  Meadow Cottage.                  Facts: hasDirectPart MeadowCottage
Ash Fleet Farm Estate has direct   Individual: AshFleetFarmEstate
  part Ash Fleet House.            Facts: hasDirectPart AshFleetHouse
```

Now asking "What are the direct parts of Ash Fleet Farm Estate?" will only return Ash Fleet Farm and Ash Fleet House, not the field or cottage. Someone is still able to ask, "What are all the components of the Estate?" by using the "has part" property in the query rather than "direct has part." This technique can be applied wherever there is a need to "switch off" the transitive restriction and will of course work for all other OWL property restrictions, such as "is connected to" as well.

10.4.5.4 Property Chains

One thing that can also be done with properties is the construction of property chains (also known as subproperty chains). To understand what these are and what they do, it is easiest to use an example. The traditional example to use is stating the relationship between an aunt and a niece (or uncle and nephew), and this is the example we use as we would hate to break with tradition. A niece is the daughter of someone's brother or sister. Without property chains, we have no way of expressing this relationship in OWL. In OWL, we can say

```
Emma has Parent Miranda.
Miranda has sister Emily.
```

And we can also say that

```
Emma has aunt Emily.
```

But really, we ought to be able to infer this rather than have to state it explicitly. This is where property chains come in. Property chains enable you to connect two or more properties and associate them to another property. The Manchester syntax to link to properties is the lowercase letter *o*. So, we can create the property hasAunt as follows:

```
hasParent o hasSister
```

We can now automatically infer the "has Aunt" relationship between Emma and Emily just by knowing that Emma's parent is Miranda, and Miranda's sister is Emily.

One place where we can apply this is to remove the problem we came across previously with the definition of Places.[12] To recap, the problem was that we had described a Property as follows:

```
Every Place is a kind of Feature.    Class:Place
Every Place has part a Feature       SubClassOf: Feature, (hasPart
 that is intended for a Purpose.      some Feature that isIntendedFor
Every Place is intended for a         some Purpose), isIntendedFor
 Purpose.                             some Purpose
```

So, we are saying that a Place has a Purpose, and that a Place has a feature that also has a Purpose. What we would like to do is enforce the fact that the Purpose of the Place overall was the same as the Purpose of is main part. We can do this with a property chain by specifying isIntendedFor as follows:

```
Everything that has a Part that    ObjectProperty: hasPart
 is intended for something will    ObjectProperty: isIntendedFor
 also be intended for that Thing.  SubPropertyChain: hasPart o
                                    isIntendedFor13
```

Once this is done, we can now describe Place as follows:

```
Every Place is a kind of Feature.     Class:Place
Every Place has part a Feature        SubClassOf: Feature, (hasPart
  that is intended for a Purpose.       some Feature that isIntendedFor
Every Place is intended for a           some Purpose), isIntendedFor
  Purpose.                              some Purpose
```

The main difference is that we now do not have to add the axiom Every Place is intended for a Purpose; the Purpose is now inferred from the Purpose of the Part.

The one downside is that as this relationship is now inferred, it is not explicit in the description of Place.

Another potential area where property chains are used is when considering the footprint of a complex object that is comprised of other features, for example, a farm that comprises the buildings, fields, farmyard, and so on. By specifying the following property chain,

```
Everything that has a Part that       ObjectProperty: hasPart
has a Footprint will also have         ObjectProperty: hasFootprint
that Footprint.                        SubPropertyChain:
                                       hasPart o hasFootprint
```

the footprint of the main feature now comprises all the footprints of the parts that make it up, so there is no need to specify a separate footprint for the main feature. Of course, this only works if we are happy to deal with the fact that we have to sum all the component footprints if we want to work out the area and total extent of the main feature, and this may introduce complexity elsewhere that we do not really want (since the ontology will not specifically tell us to sum the footprints, we will merely have inferred a Feature that has a number of different footprints belonging to it). Again, only you will know whether the use of property chains is useful for your specific situation.

10.4.6 DEALING WITH VAGUENESS AND IMPRECISION, THE PROBLEMS OF ROUGH GEOGRAPHY

Geography is not the only area that experiences vagueness and imprecision in terms of both the classes and the measurements associated with the individuals described by the classes, but it does provide some excellent examples. We shall look at how Merea Maps addresses two examples of this: imprecise distinctions between classes and handling geographic features that either do not have well-defined boundaries or where these boundaries are unknown or imprecisely known. Before discussing these examples in detail, it should be understood that because of the nature of these problems, there are no perfect solutions that completely resolve the issues. It is not a case of completely solving the problems; it is one of better managing them.

10.4.6.1 Imprecision: When Does a Stream Become a River?

The vocabularies of natural languages are littered with terms that are imprecisely defined and applied in inconsistent ways. Rivers and streams provide an excellent demonstration of this. In English, there is a general differentiation between rivers and streams since rivers are considered to be larger than streams. It is also clear that there are many instances of things named as streams that are larger than things named as rivers, and although large watercourses are always called rivers and very small watercourses are always streams, there is a blurring between large streams and small rivers. There is no established convention to distinguish the two; there are no rules to say that a watercourse above a particular length, width, or rate of flow is a river and less than that it is a stream. The first thing that Merea Maps asks of itself is, "Is it important?" Even the answer to this question is ambiguous because in one sense it really does not matter—at a physical level, there really is no difference: Both rivers and streams are watercourses; they both transport water as natural flows through channels cut through the landscape. Differences in naming are entirely a linguistic artifact. So, from this perspective it really does not matter. However, because both terms are very well established and because there is a general, although imprecise and undefined, differentiation in terms of size, it would be a strange Topographic Ontology that did not include the terms *river* and *stream* or include one but not the other. Merea Maps has a number of options. First, it can provide its own definition that precisely distinguishes between the two; for example, any watercourse with a width of 2 m or greater is a river; anything less is a stream. And, indeed this is a solution that Natural Merea arrives at when faced with a similar problem in determining the difference between lakes and ponds; here, it decides that a lake is any inland body of water greater than 2 hectares and a pond anything up to 2 hectares. However, there is a subtle difference between the two organizations' operational responsibilities that make this approach less attractive to Merea Maps. Natural Merea is using the definition for internal uses; it helps them by providing a distinction that can be used to apply different management techniques, and the choice is not completely arbitrary but founded on differences in management policies based on physical size. By contrast, Merea Maps will not use the descriptions for internal operational purposes, but as terms to enable end users ranging from professionals to the general public to query the topographic data that they are publishing. Simply imposing a size limit is unlikely to satisfy large numbers of users and could even create heated debates between individuals with very different views of what constitutes a river or stream. So, it decides to take a different approach. This approach is based on the principle that the main difference between the two is largely an imprecise linguistic distinction (or rather indistinction), and the two classes are physically identical. It therefore implements the following solution: It introduces a superclass Watercourse that represents the physical representation of a river or a stream and then creates two subclasses, River and Stream, that are distinguished by the way they are named. It then classifies instance or Rivers or Streams based on the name of the watercourse. So, the River Adder is classed as a river and the Isis Beck a stream (as becks are defined as subclasses of streams). Unnamed watercourses are deemed to be streams. This is not a complete solution. Although people can now return all rivers and streams by asking for all watercourses and can return all rivers or streams by querying the corresponding subclass, the latter could produce odd results related to the vagueness between small rivers and large streams; end users, particularly professionals, may still need a way to query by size. This is of course always possible because all the end user needs to do is execute a query for watercourse further con-strained by some physical measure of length or width. Merea Maps does one further thing to help end users; the term *watercourse* is itself less used than either river or stream, so Merea Maps provides an alternate label for watercourse—"Rivers and Streams"—that makes it clear what is meant and enables queries to be entered that use this term as well as watercourse.

10.4.6.2 Uncertain Boundaries: Places and Postcodes

There are many features that either have well-defined boundaries that are not gener-ally known or do not have well-defined boundaries at all. Features such as hospitals and universities will have very well-defined legal boundaries, but these boundaries may not be well known and in many cases will not be clearly delimited by boundary fencing. Conversely, areas such as localities within cities and postcodes in Merea do not have defined boundaries. Historically, the GI community has always been uncomfortable with these features and often attempts to impose hard boundaries on them. First, let us consider the case of a large Hospital, Medina and North Merea General Hospital, shown in Figure 10.7. It exists on one site but extends across a main road and has no physical boundary. The Hospital does have a well-defined legal boundary, but this includes a row of houses and shops that were bequeathed to the Hospital and are used by the Hospital to generate additional income. Even though these premises are therefore technically part of the Hospital, operationally they are not really related at all, and indeed most Mereans, including Hospital staff, are unaware that they are associated with the Hospital. So while the legal boundary is well defined, its definition covers more than we would think of as being the Hospital.

Merea Maps could attempt to generate an "operational boundary" that would only enclose the land associated with the health care aspects of the Hospital estate.

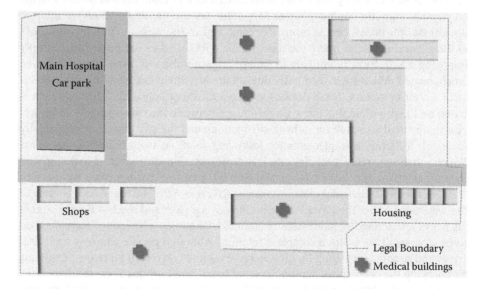

FIGURE 10.7   Hospital with no well-defined physical boundary.

Although tempting, particularly to the GI Department within Merea Maps, it decides not to do so as it feels the boundary will be an artifice that could be challenged depending on the decisions it makes to define the boundary. Rather, it realizes that what is important to most people is not a boundary, but the buildings and the access points to these buildings; people want to know which buildings belong to the Hospital, what the buildings do, and how to access the Hospital. Merea Maps therefore constructs a mereological description of the Hospital, not a geometric one.

Every Entrance is a Feature.	Class: Entrance
Every Hospital is a kind of Place.	SubClassOf: Feature
Every Hospital is intended for	Class: Hospital
Health Care.	SubClassOf: Place, isIntendedFor
Every Hospital has part a Building	some HealthCare, hasPart some
that is intended for Health Care.	(Building that isIntendedFor
Every Hospital has part an	some HealthCare), hasPart some
Entrance.	Entrance

By describing the Hospital as a Place, logic dictates that the hospital has a Footprint related to some geometry. As the actual extent is unknown, this Footprint could either be a point feature[14] that relates to a representative position, perhaps corresponding to the main hospital building (if there is one), or it could comprise all the polygons of all the buildings that make up the Hospital. Both of these are imperfect solutions; there is no perfect solution.

As we have seen, there are of course places that do not have well-defined boundaries. Frequently, localities within towns and cities are examples of these. Quite often, such places were once separate and distinct villages but over time grew together or were absorbed by a growing town or city and lost their administrative independence. As time passes, the boundaries between such localities begin to blur, so the centers are well defined but the peripheries begin to overlap, with people having differing opinions of what lies within one locality and what lies within another. Geographic Information Systems (GIS) have difficulties handling this, although a number of techniques have been developed or adopted that provide approximate solutions, such as the egg yolk method (Cohn and Gotts, 1996) or kernel density (Rosenblatt, 1956). However, both these methods rely on some degree of quantification of the vagueness at any point and so cannot be elegantly represented as Linked Data or easily described ontologically. As with the Hospital example, a mereological model of each area could be constructed. This would represent an area based on the buildings, streets, and outdoor areas, such as gardens, parks, and recreation grounds, that are known to be a part of the area. However, whereas in the Hospital example Merea Maps knew with certainty which buildings belonged to the hospital, in the locality example, we only know with certainty those Features near the center of each locality. Merea Maps handles this uncertainty by introducing a subproperty of "is part of," namely, "may be part of," and allows more than one locality to use this subproperty on the same feature as shown in Figure 10.8. Here, buildings B9–B11 and B14 are considered by some to be part of Altby and by others West Chine. A classic example of where such differences occur can be seen by examining the different views that may be presented by estate agents who are motivated by the need to present

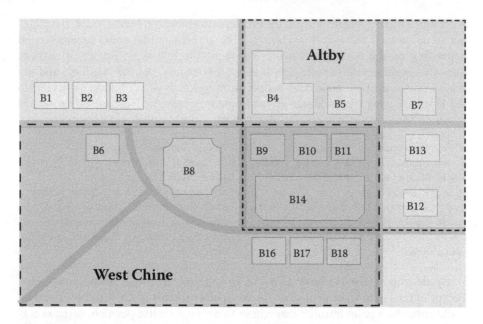

FIGURE 10.8   Areas of overlapping perceptions of West Chine and Altby.

properties as attractively as possible for sales purposes and the police who communicate with local people and who have different views of what is in which area. So,
if it is generally known that West Chine is "posher" than Altby, estate agents have
a tendency to expand the West Chine area at the expense of Altby to imply more
properties are included in the posher area. People living in the area, however, are
more likely to use different criteria, perhaps discriminating by social group, building
type, or historical continuity, and as a result this view is more likely to be reflected
by the police.

```
Altby has part B4.               Individual: Altby
Altby has part B5 (and so on)    Facts: hasPart B4
Altby may have part B9.          Individual: Altby
...                              Facts: hasPart B5
Altby may have part B14.         Individual: Altby
                                 Facts: mayHavePart B9

                                 ...

                                 Individual: Altby
                                 Facts: mayHavePart B14
```

Similarly, West Chine may also be described as maybe having B9–B11 and B14.
This allows queries to be executed that return all the features that could possibly be
part of a locality by using the "has part" property and just those that may be a part by
using the "may have part" subproperty. What this does not allow for is the ability to
ask for just the Features that are definitely part of a Locality, at least not in a straightforward manner. We can get this result by finding the inverse of the intersection of

the results of the previous two queries, but this is a little messy and certainly not clear. A more explicit way to handle this is to introduce a new subproperty of "has part," such as "has known part" (or some such) that is used to define membership where the Feature is definitely known to belong to the locality.

Mereology can therefore be used quite successfully if exact boundaries are not known. These patterns are not complete solutions, and they cannot be applied in all cases of uncertain boundaries. In some cases, the number of potential members may be so large that it makes a mereological solution either completely impractical or at best very unwieldy. In other cases, it may be simply impossible to reasonably identify the individual features, the classic case being differing types of vegetation cover that merge together, such as rough grassland merging into scrub, which in turn merges into woodland. In both these cases, Merea Maps has little option but to resort to boundary estimations expressed as overlapping polygons.

10.4.6.3   Working with Insufficient Data

There are some things that are very difficult, if not impossible, to define using OWL. Consider the following two examples encountered by Merea Maps: The first is a braided river. Braided rivers are defined as rivers that have at least one stretch that contains multiple channels.[15] We can immediately state the subclass relationship:

```
Every Braided River is a kind of River.    Class: BraidedRiver
                                           SubClassOf: River
```

But after that, it becomes difficult. The following statements superficially appear to meet the second requirement for braidedness:

```
Every Braided River has part a        Class: BraidedRiver
 Braided Stretch.                     SubClassOf: hasPart some
Every Braided Stretch has at least     BraidedStretch
 two Channels.                        Class: BraidedStretch
                                      SubClassOf: hasChannels min
                                       2 Channel
```

The problem is in the way we interpret "multiple channels." Strictly, any river stretch that has more than one channel has multiple channels, hence the previous definition that follows this strict interpretation of multiplicity. And, from a mathematical viewpoint, and even a strict linguistic viewpoint, who could argue? The problem is that no one would describe a river stretch with just two channels as braided. The definition uses "multiple" quite loosely, and this reflects the lack of real definition of braided rivers, or more to the point that it is difficult to answer the question: How many channels are required to make it braided? No one has ever precisely defined it: Not two, probably not three or four either. What about five? But, because no one has defined the exact number, then the question is impossible to answer. People are able to recognize braided rivers when they see them, so it is possible to classify them. It can only be done in a precise mechanistic way with great effort, and that has simply not been done for the great majority of braided rivers. Merea Maps therefore has three options. It can use some vague and undefined property "has multiple":

```
Every Braided Stretch has          Class:BraidedStretch
  multiple Channels.               SubClassOf: hasMultiple some Channel
```

It can set some lower limit that it believes most people will agree with:

```
Every Braided Stretch has at       Class:BraidedStretch
  least 5 Channels.                SubClassOf: hasChannel min 5 Channel
```

Or, it can say nothing about the number of channels at all. Saying nothing is not quite doing nothing; Merea Maps can still subclass the Braided Rivers and classify rivers as braided, and users can ask questions about them. This is an important example of using the open world assumption to advantage—a precise definition cannot be produced, so do not try; rather, define a class and let experts on the ground apply the class. If they record the number of channels, then over time it might even be possible to deduce a minimum. There will be many occasions when you can spend a lot of time trying to find precise definitions that simply cannot be nailed down. The important thing is to recognize early on the nature of the beast and to quickly come to terms with an incomplete description.

10.4.7   DEFINED CLASSES

We have been careful when talking about the specification of a class to say that we are *describing* rather than *defining* it. This is because there is a way of specifying a class that specifically results in what are termed *defined classes*. A defined class differs from a described class in that it enables a reasoner to automatically classify any individual as a member of the class providing that individual meets the criteria of the defined class. To understand how a defined class works, you need to understand necessary and sufficient conditions.

10.4.7.1   Necessary and Sufficient Conditions

To this point, we have been dealing with what are called primitive classes, classes that are simply subsets of other classes. In DL terms, if A is a subset of B we say that B is a necessary condition for A. If we look at geographic examples, then for a river to be a river, it is necessary for it to be a type of watercourse, so being a watercourse is a *necessary condition* for being a river; for a schoolhouse to be a schoolhouse, it is *necessary* for it to be a building and for it to have the purpose of providing education, so being a building and providing education are *necessary conditions* for being a schoolhouse.

  If we look at the problem from the other direction, then we can see that if we know something is a river, then it must also be a watercourse. In this case, we say that a river is a *sufficient condition* for something to be a watercourse: If something is a river, it must also be a watercourse. But equally, being a river is not a necessary condition for being a watercourse since a canal or drain is also a watercourse but are not a river.

Defined classes are classes whose restrictions on properties are both necessary and sufficient conditions. What this means is that anything having that combination of properties must be the defined class. This is best expressed by the Rabbit statement "Anything that." Consider the following:

```
A Wood is anything that:          Class: Wood
is a kind of Landform;            EquivalentTo: Landform and
   has dense cover of Trees.        hasDenseCover some Trees
```

What this is saying is that any piece of land that has dense tree cover must be a wood. In terms of necessary and sufficient conditions then, we can say that to be a wood it is necessary for it to be a piece of land and for the land to be densely covered by trees, and that being land that is densely covered by trees is sufficient to be a wood. Rather than merely describing the class, we have defined it. Another way to look at this is that the class Wood is equivalent to the class of things that are land with dense tree cover.

As we can see, the difference in OWL between a Defined Class and one that is merely described is that the equivalence rather than subclass relationship is used. Defined classes are quite powerful because a reasoner will classify any individual that meets the necessary and sufficient conditions as being a member of the defined class, irrespective of whether we have explicitly stated its membership. For example, any other thing that is defined as land and has dense tree cover would be automatically classified by a reasoner as a wood. This can be very useful, but equally it is very easy to get it wrong; we have to be really sure that what we say is true. As an example, consider a Bakery. We might want to define a Bakery as follows:

```
A Bakery is anything that:        Class: Bakery
   Is a kind of Place;            EquivalentTo: Place that
   Is intended for Baking of Bread.  isIntendedFor BakingOfBread
```

So, any place that has a purpose of baking bread is a bakery, a statement that sounds quite reasonable. However, consider a large supermarket that also bakes its own bread:

```
Every Supermarket is a kind of Place.  Class: Supermarket
Every Supermarket is intended for the  SubClassOf: Place that
  Sale of Groceries.                     isIntendedFor SaleOfGroceries
Medina Merea Superstores is a          Individual:
  Supermarket.                           MedinaMereaSuperstores
Medina Merea SuperStores is intended   Types: Supermarket
  for Baking of Bread.                 Facts: isIntendedFor some
                                         BakingOfBread
```

What this says is that the particular shop Medina Merea SuperStores (the Medina Branch of the Merea SuperStores chain) is a supermarket, and it also bakes bread. From the fact that it is a supermarket, we know it is also a place. This means that it has both the necessary and sufficient conditions also to be classified as a bakery (the fact that it has additional conditions is irrelevant). For some people this might be okay, but many people would not say it was a bakery but a superstore that bakes bread. The difference is subtle but depending on viewpoints may be either right or wrong. The message is to be careful when wanting to make something a defined class because it can have unforeseen effects elsewhere. So, whereas these conditions are not enough to define a bakery, there are usually things that can be done to make many classes defined rather than described. In the case of the bakery, we could describe a new property "has primary purpose" that is a subproperty of "is intended for" and use this to redefine the bakery class:

```
A Bakery is anything that:          Class: Bakery
   Is a kind of Place;              EquivalentTo: Place that
   Has primary purpose Baking of Bread.   hasPrimaryPurpose
                                    BakingOfBread
```

Now, things such as supermarkets that may bake bread but not as a primary function will not be incorrectly classified as bakeries.

We can also go back and reconsider other classes that could be candidates for being made into defined classes. One possible example is Orchard, discussed previously:

```
Every Orchard is a kind of Enclosed Land.   Class: Orchard
Every Orchard has part Trees.               SubClassOf: EnclosedLand,
Every Orchard is intended for                  hasPart some Tree,
   Agricultural Production.                     isIntendedFor some
                                               AgriculturalProduction
```

If we can satisfy ourselves that *anything* that is enclosed land, contains trees, and is intended for agricultural production can be considered to be an orchard, then it is reasonable to *define* the Orchard class:

```
An Orchard is anything that:        Class: Orchard
   is a kind of Enclosed Land.      EquivalentTo: EnclosedLand,
   has part Trees.                     hasPart some Tree,
   is intended for Agricultural        isIntendedFor some
Production.                            AgriculturalProduction
```

10.4.8 DIFFERENT PERSPECTIVES: LAND COVER

Quite often, different people will have different perspectives of the same things. Most usually and most starkly, this can be seen between different organizations where the perspectives are defined by the differing operational needs of those organizations; the Merean government's taxation agency will have a very different perspective on a building than that of Merea Heritage. But, such differences can also occur with a single organization and indeed within a single ontology. As an example, we consider the approach taken by Merea Maps with respect to the relationship between two hierarchies, one describing Features and the other Land cover. Land cover is a form of classification that provides a means of summarizing what exists on the land's surface. It is useful for ecological and environmental purposes and even helps in

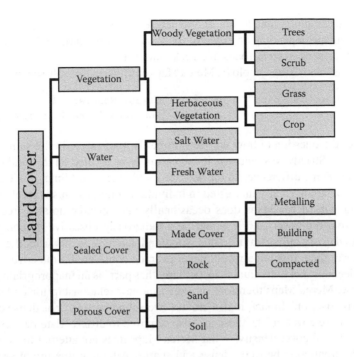

FIGURE 10.9 Landcover classes.

the production of traditional cartographic maps, enabling areas such as woodland and marsh to be clearly shown and differentiated on mapping. Most existing hierarchies can be a bit confused in places, often mixing land cover with land use—terms that, although related, are clearly different; the former deals with what is on the surface and the second how the land is used. But, there is also confusion between what something is—the type or class of a Feature—and what is *on* it or part of it. For example, a typical land cover class is Woodland, but is this not really a type of Feature? Strictly, from a land cover perspective, the cover is trees, not woodland, but in common parlance people have conflated the two such that it would not seem odd to many people that woodland was considered to be a land cover class. What is certainly true is that what makes something a woodland are the trees that cover the land—trees can be seen as a part of woodland just as buildings can be seen as being part of a hospital or school. These observations enable Merea Maps to do two things; first, it is able develop a land cover ontology that is more truly about land cover, and it is then able to use this land cover ontology to help define its topographical ontology (Figure 10.9).

If Merea Maps were focused exclusively on topographic definitions, then it could get away with just incorporating land cover aspects into its Topographic Ontology. Merea Maps, though, decides that if it produces a separate Land Cover Ontology that is then reused by the Topographic Ontology, this will provide its end users with greater flexibility. This decision also reflects the difference between the upper levels and purposes of the two ontologies. The Topographic Ontology is more concerned with differentiating classes on broad topographic groups such as settlements, administrative regions, land forms, and structures, whereas the Land Cover Ontology will differentiate depending on physical form and will discriminate between vegetation, solid surfaces (such as roads and rock), and water.

Using this Land Cover Ontology, Merea Maps can now reasonably say the following:

```
Every Woodland has part Trees.        Class: Woodland
                                      SubClassOf: hasPart some Tree
```

This raises the question of how we view a tree: Is it a Feature in the same way that a building is? Strictly, yes, because it can certainly be located on the land's surface, but there are some differences in the way we should view Buildings and Trees. With buildings, we are normally interested in individual buildings; this is rarely true with trees. Although Merea Maps does occasionally map specific notable trees, this is very far from the norm, and it would be a ludicrously obsessive mapping agency that mapped every individual tree in a woodland. Coming at this from the direction of land cover then, this issue would never even be considered—the cover is always about collections, not individuals. So, perhaps "has part" is an inappropriate property in this case. Merea Maps therefore introduces a new relationship for its land cover types: "has cover of." In fact, it also creates two subproperties, "has dense cover of" and has "sparse cover of," as these are also required to differentiate between different types of land cover density. What Merea Maps does not attempt to do is state a precise differentiation between dense and sparse. Although this might seem to be introducing imprecision, it reflects the fact that it is almost impossible to be precise simply because only rarely do people go to the trouble of accurately measuring the density of tree cover. Normally, estimation is based on the eye, and crude differentiation between dense and sparse is sufficient. So, what may be seen as imprecision in the ontology is actually just reflecting normal working practices and a pragmatic acceptance of something that is fit for its purpose. Now, Merea Maps states:

```
Every Woodland is a kind of           Class: Woodland
  Landform.¹⁶                         SubClassOf: Landform
Every Woodland has dense cover of     Facts: hasDenseCoverOf some
  Trees.                                Tree.
```

These new properties can be used to help describe individual instances of Feature types. For example, Isis Heath can be described in the following terms:

```
Every Heath is a kind of Unenclosed   Class: Heath
  Land.¹⁷                             SubClassOf: UnenclosedLand that
Every Heath has dense cover of          hasDenseCoverOf some Heather
  Heather.                            Individual: IsisHeath
Isis Heath is a Heath.                Facts: hasSparseCoverOf some
Isis Heath has sparse cover of Trees.   Tree, hasSparseCoverOf some
Isis Heath has sparse cover of          Boulder
  Boulders.
```

This tells us that Isis Heath not only has a dense covering of heather (a necessary condition of a heath) but also has a scattering of trees and boulders.

Merea Maps also notes that there is also a new pattern emerging: that of Feature + Cover = New Feature. So, for example, we can see that:

Every Heath is a kind of Unenclosed Land. Every Heath has dense cover of Heather.	Class: Heath SubClassOf: UnenclosedLand that hasDenseCoverOf some Heather

In some cases, the addition of a function is required. An example of this is an orchard, which is a constructed feature with a clear purpose, which helps to define what it is:

An Orchard is anything that:
    is a kind of Enclosed Land.
    has part Trees.
    is intended for Agricultural
Production.
	Class: Orchard
	EquivalentTo: EnclosedLand,
	  hasPart some Tree, isIntendedFor
	  some AgriculturalProduction

In contrast, a Heath has no clear purpose or reason for being: It just is.

10.5   ONTOLOGY REUSE: AIDING THIRD-PARTY DATA INTEGRATION

So far, we have only discussed combining ontologies in a fairly simple fashion. To this point, we have always assumed they are additive; that is, they provide use with extra classes, predicates, and axioms that we are able to use directly within our ontology. Indeed, the theory behind micro-ontologies is that they are just this: additive. But, what about situations where the ontologies are more complex than micro-ontologies and there are semantic differences between them that cannot be resolved easily because the ontology author does not have control or authorship over all the ontologies? This is the type of problem that will be faced by any organization wishing to use the data supplied by Merea Maps and described by its Topographic Ontology. The ontology is complex and clearly not a micro-ontology. It describes the world as seen by Merea Maps, and although the data and the ontology are recognized by a third party—say, Merean Nature—as useful, Merean Nature also recognizes that it has a slightly different worldview that means it will not always be in total agreement with Merea Maps. Nonetheless, the overall benefit of using the data and ontology from Merean Maps outweighs the perceived issues, so Merean Nature decides to use the Merea Maps' data in its business processes and the ontology to describe it.

One task that Merean Nature has is the need to provide an annual health report on the state of the Merean environment in terms of the well-being of the various habitats that exist on the island. These health reports are used to both inform central government and to form the basis for action plans agreed between Merean Nature and the various local authorities to ensure the habitats are properly managed. To do this, it needs a topographic view of the world, but this view itself is not sufficient to enable Merean Nature simply to deduce the habitats. Its plan is to copy the example of Merea Maps and to produce its health plan as Linked Data described by an ontology.

It believes this will enable its maximal reuse of its data and maximal use by central and local government and by the environmental and construction companies that also use its data.

Habitats are usually very closely associated with land cover, so many of the features and land cover classifications applied by Merean Maps are of immediate use to Merean Nature. Merean Nature has already produced an ontology that describes the habitat classification system that it uses and so begins the process by trying to match the land cover ontology provided by Merea Maps with this ontology[18]; this process is known as ontology alignment. What alignment attempts to do is match classes in one ontology with corresponding ones in the other ontology. This is similar to the process for Link Discovery (in particular, vocabulary links) outlined in Section 8.6, with the added complexity of descriptions being in OWL rather than the simpler RDFS.

For example, both Merea Maps and Merean Nature have a class Woodland, so if it can be demonstrated that these two classes are describing the same real-world objects, then we can link them together using the predicate `owl:equivalentClass`. This predicate is similar to the `owl:sameAs` in that it is to classes what `owl:sameAs` is to individuals—it states that they are exactly the same thing. That is, every statement that has been made about one class is also valid for the other class. And, like `owl:sameAs`, it is very easy to misuse and associate two classes that are not quite the same thing, so care must be taken.

Ontology alignment is still the subject of much research; a significant amount of this is directed at trying to automate this process (Choi, Song, and Han, 2006; Noy and Stuckenschmidt 2005, Bouquet, 2007). It is admitted by most that at present the best that can be achieved is a semiautomated solution with either a human expert being required to correct an automated process, or the process being used to assist the expert by suggesting possible alignments between ontologies.

Merean Nature decides to err on the side of caution and align the ontologies using its experts and does not attempt to use a semiautomated solution. Figure 10.10 shows an extract of the habitat class hierarchy that Merean Nature has constructed, which it has based on the JNCC Phase 1 Habitat system. The system is similar to the land cover system devised by Merea Maps, but it is not the same. If we look at Woodland Habitat, one of the properties it has is that the Woodland Habitat is contained within Woodland.

```
Every Woodland Habitat is only        Class: WoodlandHabitat
  contained within Woodland.          isContainedWithin: some Woodland
                                        and
                                      isContainedWithin: only Woodland[19]
```

As indicated, Merean Maps wants to link its concept of woodland to the woodland concept in Merea Maps' ontology. This will allow it to infer that woodland habitats exist wherever Merea Maps has mapped woodland. By examining the full definition of woodland, as specified by Merea Maps, the Merean Nature domain expert is able to determine that they are the same classes and can therefore be linked together[20]:

```
Woodland and Woodland [Merea Maps]    Class: Woodland
  are equivalent.                     EquivalentTo: mm:Woodland[21]
```

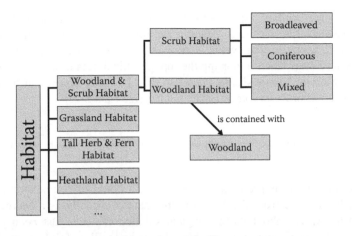

FIGURE 10.10   Part of Merean Nature's Class hierarchy for Habitats.

As indicated, this is a very strong statement to make. It means that the two classes are the same class, so anything that is classified as a woodland by Merea Maps would also be classified as such by Merean Nature and vice versa. Also, all the statements made in the ontology about Merea Maps' Woodland are also true for Merean Nature's Woodland.

Both Merea Maps and Merean Nature make the assertion that Woodlands contain trees, but they define the class hierarchies leading to trees differently. In the case of Merea Maps, it has applied the principle that from a topological point of view a tree is a secondary concept, so it has said nothing about it and made it a direct subclass of Thing. Merean Nature, on the other hand, does know a bit about trees and so has a more complex class hierarchy that can be summarized as Thing, Vegetation, Woody Vegetation, Tree. On the face of it, then, these classes are different; however, because there is actually no conflict, Merean Nature can also make these classes equivalent as well. This is due to the fact that no contradictions exist: In both cases, a Tree is a subclass of Thing, and all that the Merean Nature's hierarchy does is add more specificity. As important, Merean Nature's domain expert is happy to accept that Merea Maps is using the concept of tree in the same manner as Merean Nature. It is also worth considering that if the domain expert had decided that Merea Maps was using the concept of tree differently, then Merean Nature would probably also come to the conclusion that the Woodland classes were also not equivalent as both refer to their respective Tree classes.

An example of where differences certainly occur can be seen if we consider what happens when the domain expert tries to address whether Lake is used in the same way in both ontologies. In the Merea Maps ontology, a Lake refers to an open area of water that has a significant flow; Merean Nature describes a Lake as any area of open water larger than 2 hectares. There is clearly a relationship between the two, but they are also not equivalent: There will be certain individual Lakes defined by Merean Nature that would not be classified as Lake by Merea Maps and vice versa. However, the domain expert notes that anything that Merea Maps has classified as a pond, reservoir, or lake that has a surface area of greater than 2 hectares will be classified as a lake by Merean Nature. The problem is that OWL does not enable someone to specify the "greater than" aspect, so what it does is create a value "LakeSize" and property hasSize that is used to denominate any pond, reservoir, and lake that is greater than 2 hectares. This value is assigned as a part of the preprocessing that Merean Nature does when importing the topographical data it receives from Merea Maps. The Merean Nature's Lake class can then be made equivalent to an anonymous class (Pond or Reservoir or Lake) and hasSize LakeSize:

```
Class: Lake
EquivalentTo: (mm:Pond or mm:Reservoir or mm:Lake) that hasSize value
LakeSize[22]
```

As Merean Nature is preprocessing the data, it could of course simply classify every eligible feature from the Merea Maps' dataset directly as a Merean Nature Lake and not worry about the ontological solution that has just been presented. On the face of it, this seems a reasonable solution and will certainly work. However, the downside is that this hides the relationship in code rather than making it explicit, which is one of the major points of using an ontology.

There will of course be areas where nothing can be done, and no relationship can be reasonably constructed between two classes, for example, if the external dataset lacks the data needed to extract such a relationship and its ontology can provide no corresponding description. As a case in point, Merean Nature has a habitat class Still Anoxic Freshwater. Such a habitat can occur in things that Merea Maps has classified as Pools, Lakes, Reservoirs, Canals, and Rivers and Streams, but only where there is no flow and the oxygen levels are very low. Merea Maps simply does not record oxygen levels, and still waters can occur in parts of all these freshwater features, but again other than for Ponds, Merea Maps does not record where still waters occur. So, the reality is that there is little that Merean Nature can do in this situation as there is no useful Merea Maps class to which they can link.

One final thing to consider is: Where class equivalence is known to be valid, is it better to create a local class and make it equivalent to the external class, or is it better to just use the external class? If an area as young as ontology authoring can be said to have a tradition, then it has been traditional to say that the preferred solution is always to reuse the external class rather than also creating a local class. However, there is a downside to this because more and more ontologies need to be referenced the further down the reuse chain you go, and the system can become unmanageable. So, there is an argument to say that if you in turn expect your ontology to be reused by others, then it may be more sensible to create local classes and make them equivalent in a separate ontology file so that others who wish to reuse your ontology, but are not necessarily interested in other ontologies, can do so without pulling in too many unnecessary ontologies. Figure 10.11 shows how this works. The Main Local Ontology contains the body of the ontology, including local descriptions for the classes that are also present in the External Source Ontology. In the case of Merean Nature, the Main Local Ontology would contain all its habitat descriptions along with local classes for things such as Woodland. The Merea Maps ontology is the External Source Ontology and also contains a description of Woodland. These classes are linked by a separate Merean Nature ontology that just contains

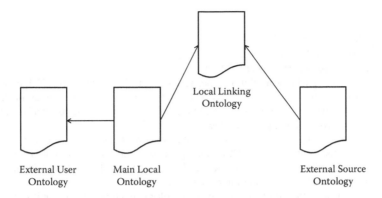

FIGURE 10.11   Modularizing ontologies to reduce unnecessary reuse.

equivalence statements between the two main ontologies. (This has parallels with Linked Data Linksets described in Section 7.6.) A third party wishing to use only the Merean Nature ontology can now do so as it only has to refer to the main Merean Nature ontology. If it does also want to use the Merea Maps ontology, it can do so by using the linking ontology.

Of course, another case where it may be sensible to have local classes is when two preexisting ontologies are linked, so dropping local classes in favor of the external ones is not cost effective.

10.6   SUMMARY

This chapter has established the basic principles of constructing a geographic ontology and has shown how these can be used to aid data integration. A number of methodologies exist to guide the construction of an ontology; almost all of the emphasis is placed on the importance of establishing a well-defined scope and purpose for the ontology that helps to guide the author. Most of these methodologies also stress the importance of establishing competency questions early on that can be used to test the correctness, completeness, coverage, and interlinking of the ontology. The construction process itself comprises iterative phases that involve the development of a lexicon and glossary that then helps guide the description of the classes and properties of the ontology. To those not familiar with the authoring of an ontology, the Open World Assumption may require a little acclimatization, as will the fact that properties can be specialized. The geographic domain itself offers the ontology author a number of challenges. Ontologies are well suited to expressing network topology and mereological relations, and reusable modeling patterns also begin to emerge where features are described in terms of their physical components, function, or land cover. However, not all geographical knowledge can be fully expressed in DL, so an understanding of the limitations of OWL is important.

In the next and final chapter, we present a summary of the book, discuss what may be on the horizon for GI as the Semantic Web develops, and emphasize for a last time the most important aspects of Linked Data and the Semantic Web when applied to GI.

NOTES

1. http://www.opencyc.org/.
2. http://www.ontologyportal.org/.
3. http://www.loa.istc.cnr.it/DOLCE.html
4. Note even though this is an RDFS property, it is still available for use by OWL ontologies.
5. Obviously, in the case where the two terms have been stated as classes in different ontologies, then we would need to make them equivalent classes, but this is something that can be avoided within an ontology.
6. This is the short form of: http://www.opengis.net/rdf#. The OGC geometry ontology also contains vocabulary for RCC8 topology as well, but in this example we are showing these described in a different imaginary ontology to emphasize how different aspects can be obtained from different micro-ontologies.
7. This emphasizes the difference between mereology and topology; while it is not unreasonable to exclude a steelworks from being a part of a farm—a mereological exclusion—it is possible that a steelworks could be contained within the extent of a farm, which is a topological possibility.
8. Note that in Manchester syntax "and" and "that" are used interchangeably. It is the brackets in the sentence that make it differ from the previous example.
9. Assume here that Enclosed Land is a kind of Landform.
10. It is also worth remembering at this point that whereas subproperties inherit range and domain restrictions, they do not inherit any OWL restrictions. Although this is somewhat odd, it can be turned to advantage, as we shall see.
11. It should also be remembered that the inverse property will also not be inherited, so this will have to be explicitly stated for "has direct part."
12. For now, let us forget about the modification we made to Place by making it a subclass of Site or Building and just concentrate on the key problem we identified before making the modification.
13. This definition might look a bit recursive, but in fact it is not. What the statement says is that if the pattern hasPart o isIntendedFor occurs, this will cause the reasoner to infer the isIntendedFor property exists.
14. Specifying that a Footprint can be represented by a single point rather than by an area may seem odd or more bluntly wrong. However, the key here is the word *representative*: The hospital obviously has an area associated with its Footprint, and the point is just used to represent this area in the absence of more precise knowledge about it.
15. In fact, another more scientific definition exists based on gradient and deposition rates and dependent on the sediment material (Schumm and Kahn, 1972). This definition is even more convoluted and only presents the same problems. Let's not go there!
16. This is one of the top-level classes of Feature.
17. In turn, this is a kind of Landform.
18. Although habitat and land cover classifications are different, they are nonetheless related as land cover is a major component of many habitats.
19. Note that in Rabbit "only" means that something must have that property value and can only have that class or value, whereas in Manchester Syntax to achieve this you have to use both "some" and "only."
20. Alternatively, Merean Nature could decide simply to replace its woodland class with the Merean Maps' class.
21. This is where mm: expands to https://ontology.mereamap.gov.me/.
22. This is only shown using Manchester syntax as anonymously classes cannot be expressed in Rabbit.

11 Linking It All Together

11.1 INTRODUCTION

In this chapter, we provide a summary of the book and highlight those points we think are especially important, along with some suggestions for directions that the Semantic Web, and the position of Geographic Information (GI) within it, may take in the future. We believe that Linked Data and the Semantic Web will have an increasing impact on the manner in which data is managed and utilized. It will in turn influence and affect the way in which GI is treated and has the capability to make the use of GI much more widespread. The technology behind the Semantic Web is itself far from perfect and in places still quite immature. We have not tried to present it as a panacea, and it certainly cannot always model GI as accurately as we might wish; this limitation also extends of course to any other kind of information. But, by understanding the limitations inherent in both the technology and the GI (or more generally, the data) that we have, we can still develop useful models of our knowledge and build applications based on these models.

11.2 THE WIDE SCOPE OF GEOGRAPHIC INFORMATION

One of the first things to remind ourselves of is that GI has a much broader definition than the data we may find within a GIS. It is certainly true that most of the thought on GI, and almost all of the standards, has come from the community of Geographic Information Systems (GIS) users, broadened slightly by those interested in database technology such as Oracle. It should be equally clear, however, that very significant amounts of data that have geographic components exist outside this community.

While the usage of GI may not be quite as ubiquitous as some of those within the inner circle may believe, it is nonetheless very widespread; location is a powerful common and shared thread that runs through many data sources. Where such threads intertwine between datasets, shared location can therefore act as an important binding agent between data.

However, conventional means of representing and manipulating GI do not always make it easy for data to be exchanged and integrated. Where standards for exchange exist, they tend to be geometry and geography centric—they largely focus on the geometric component and stem from the belief that the user is fundamentally interested in the geography included in the data representation. This is certainly true of the GIS specialist and for the many others who do regard geography and geometry as of inherent importance; for them, these standards are a good solution. However, other people may well find the Open Geospatial Consortium (OGC) standards less useful because geography is less important for them, and their primary focus is elsewhere. As there is a good chance they do not use GIS, they are probably unfamiliar with OGC standards, and their interaction with data is dominated by other standards.

225

These people cannot of course be completely neutral or ambivalent to geography if they wish to use it for data integration. And, it is here that Linked Data and the Semantic Web may help by providing a common and simple data model based on the Resource Description Framework (RDF) triple and a machine-interpretable means to describe the data using ontologies. This common data model provides a foundation that simplifies the process of data integration or linking and is also data neutral: The same standard applies to bioinformatics just as much as to GI or financial data. By separating the description of the data from the data itself and exposing it in an ontology, the Semantic Web approach also allows a third party to understand the meaning of the data better and to begin an integration or linking process at this level. The establishment of links between data enables relationships to be explicit and visible. Hence, the decisions that have been made during the data integration process are made clearer and are also preserved for others to use.

For the GI community, the adoption of these technologies offers the chance to open up their expertise and specialisms to a much wider audience. The manner in which the spatial elements of data are represented and manipulated will be preserved, often simply by finding ways to represent existing standards (or the relevant aspects of them) within a Semantic Web environment. And, once preserved in this new environment, they naturally become open to many more end users. The general way in which data is formatted and exchanged now exists within a Linked Data framework, and the specialist aspects of the data, such as the representation of geometry, can be enshrined as special datatypes described by the GI community. Any person who uses Linked Data, irrespective of the person's background, now has a common representational form for his or her data that will be familiar to any other person who also uses Linked Data. The GI community can also publish micro-ontologies that define and describe the vocabularies that are used, in a manner that can be understood by a much wider audience than the GI community alone. A major barrier to data integration, the problem of varied formats, has therefore been significantly reduced. The advent of Linked Data now means that GI really does belong to a much broader community.

GI itself is evolving; there is increasing recognition of what can be pithily summarized as "Place before Space." By this we mean that a lot of problems can be resolved without knowing very precise locations or extents as would be traditionally required by a GIS. Place is more concerned with identity, which includes place names, addresses and postcodes or zip codes, and topologic and mereologic relationships. Location itself is often expressed as a simple point, and the boundary of an object may not be represented at all.[1] Such emphases are well suited to expression as Linked Data. We have also seen that some aspects of GI are not well suited to explicit representation and analysis on the Semantic Web. For example, raster data can only be referenced by a Uniform Resource Identifier (URI) and cannot be interacted with at all on the Linked Data Web, while analysis and manipulation of vector geometries is quite limited. This means that GIS will therefore remain an important analytical tool, standing alongside the representation of data on the Semantic Web. Most benefit will therefore be realized by recognizing that the technologies are largely complementary rather than competitive: GIS are better suited to analysis and to raster and vector data, while the Semantic Web is better at the representation of data in a form that maximizes reuse, interaction, and integration.

The following are points to remember:

- GI is really a broad church and exists in many different forms.
- GI is an important element in the integration of data;=.
- GI standards can be encapsulated within the Semantic Web with vocabularies described by micro-ontologies and standardized geometric representations expressed as special datatypes.
- GI itself is evolving, and great recognition is being given to "Place." Place is well suited to representation as Linked Data.
- Linked Data and the Semantic Web can help GI become a more integrated part of the wider information community.
- Linked Data and the Semantic Web do not replace the need for GIS. They cannot match the analytical properties of a GIS, and they are not well suited to all forms of geographic data, such as raster. Many aspects of the older GIS technologies are therefore complimentary to the Linked Data approach.

11.3   AN OPEN WORLD

For those not previously aware of the open world assumption, the implications of its application are one of the most important messages in this book. The open world assumption pre-dates the Web, but its way of looking at the world is ideally suited to dealing with data held on the Web. The crisp boundaries that define the scope of a conventional database are simply not present on the Web, and we therefore cannot assume that if we cannot find a fact it cannot be true and so must be false. The Web itself is also littered with contradictions: "facts" posted by different publishers that differ from each other. This world we experience on the Web is also found in everyday life. Conventional databases essentially ignore these difficulties to provide a consistent and managed view of a problem domain. This strategy is entirely sensible, especially where there is an internal focus within an organization or across a closed group of like-minded organizations. The strategy is less successful when applied to a broader population, where a diversity of perspectives exists, that cannot, for legitimate reasons, be coerced into a single shared worldview. Indeed, a characteristic of the Web, and life in general, is that there is no one single worldview. On the Web, strategies based on the open world Assumption fare better as the goal is no longer to share a single view but to take part in a discourse among many different viewpoints.

One way in which this difference manifests itself is the way in which we deal with information that we had not considered at the outset. In the closed world approach, we are obliged to determine in advance all the types of data that are of interest. If we are dealing with a road, then perhaps we will identify that we need to hold its name, the identities of the roads to which it connects, and its position and geometry. We may also create a constraint: that all roads must have a name. Once done with modeling, we can go about populating the database with values representing the roads in which we are interested. And, we can of course do the same when applying the open world assumption and representing the information as Linked Data.

The difference comes when we wish to add a road but do not know the name or realize that for another road it is also important to store the road's old name as well as its current name. The closed world solution is unable to handle either of these cases without changing the database schema and constraints. We cannot store a road unless it has a name, so if we wish to store a road where the name is unknown, then we have to relax the name constraint. In the case of adding new data, we must physically alter the database structure to allow this new data to be held. In an open world, neither challenge is problematic. We are able to enter the road with the unknown name as the constraint just tells the system that the road has a name; we just do not know what it is. Adding a new item of information is also straightforward: a new predicate say, has _ old _ name is created and used to reference the road's previous name. No change is required to the data structure; the structure was triples before and remains triples afterward.

Of course, there are times when we want a closed world, especially if we are dealing with data internal to a single organization. Here, an open world approach can be more than a little frustrating. From an initial purist standpoint, there is increasing recognition that there are times when it is useful to "switch off" open world reasoning and allow standard database constraint rules to take over. In the future, it is therefore likely that tools will increasingly enable users to toggle between closed and open world approaches. Initial mechanisms for this are appearing in tools like Topbraid SPIN, discussed in Chapter 8, which harnesses SPARQL queries to test for Linked Data validity.

The following are points to remember:

- The open world assumption is a very different way of viewing information.
- The open world assumption allows us to cope with the boundless extent of the Web.
- It differs from a closed world solution by requiring us only to specify minimum characteristics of things, not everything we think we need to know. This means we can vary the information we hold about specific feature types, we can add more things about specific instances, and we can work with incomplete data.
- Sometimes, it is useful to enforce closed world methods, particularly when we want to engage integrity checks on internal data.

11.4   THE SIMPLICITY AND COMPLEXITY OF THE SEMANTIC WEB

One of the great strengths of Linked Data is its data structure and general simplicity. Once people have understood the idea behind triples, the concept of Linked Data as a graph or network of interconnecting triples is very easy to understand. Those coming to RDF from an XML (eXtensible Markup Language) background will have to look past the tags to see the graph, but the basic concept is pretty easy once it has been grasped. The great elegance with this solution is that this data structure is both simple and universal: All Linked Data is expressed as triples of the form *subject predicate object*. Subjects may represent classes of things or individuals, predicates (also known as properties or relationships) establish the link between subject and object. Objects may in turn be classes of things or individuals or may be represented by a value (also known as a literal).

A very significant advantage of the triple data structure is that one of the current bugbears of data integration, that of discovery and manipulating diverse data structures that are often poorly documented, simply goes away; all data on the Linked Data Web has the same structure: the triple.

Linked Data is also built on well-known and proven technologies such as HTTP (Hypertext Transfer Protocol). Everything is identified using HTTP URIs, meaning that it is possible to mint identifiers such that their uniqueness can be ensured. By making the URIs dereferenceable, they can be used to point to the data that underlies the concept, individual, or property in question. Hence, the Linked Data Web can be seen as a network of nodes made up of URIs representing classes and individuals, which are linked by URIs relating to properties. Publishers are able to build links between their data and the data published by others simply by adding triples that establish the links.

Data on the Linked Data Web can be described using ontologies; these may be simple RDFS (RDF Schema) ontologies that do little more than specify vocabularies used by the Linked Data, or they may be more complex and expressed in OWL (Web Ontology Language), thus enabling sophisticated inferences to be made over the data. An advantage of describing Linked Data using ontologies is that they remain independent of the application code, so the meaning of the data is much more visible than in conventional solutions, where much of the structure is buried in the application code.

The development of OWL ontologies can be quite a complex process, and a number of methodologies exist to assist the process. As OWL is based on first-order logic, it can be quite difficult for nonlogicians to become familiar with the more subtle aspects. It is also easy to confuse OWL classes with classes from the object-oriented (OO) paradigm as they look very similar on the surface.

The following are points to remember:

- The basic and universal structure of Linked Data is the triple, which represents the relationship between a subject and object associated by a predicate and expressed as *subject predicate object*.
- The Linked Data Web uses well-proven technologies such as HTTP and the URI scheme.
- Everything is identified using URIs.
- Publishers are able to build links between their data and the data published by others simply by adding triples that establish the links.
- All things other than data values are identified by URIs, and these URIs should deference to return the data that is associated with the URIs.
- Linked Data may be described using ontologies. The ontologies themselves may be very simple (often expressed using RDFS) and may be little more than specifications of the vocabularies used to describe the Linked Data, or they may be more complex, expressed using OWL, to enable inferences to be made from the Linked Data.

- Ontologies are independent of the implementing application and code, increasing the visibility of the descriptions they provide of the data.
- Developing OWL ontologies can be difficult, and methodologies exist to aid the process.
- OWL is based on first-order logic, and it can be difficult for nonlogicians to fully understand all the logical consequences of its reasoning.
- OWL classes are very different animals from OO classes; it is easy for the novice with experience in OO design or programming to confuse the two. Do so at your peril.

11.5   THE TECHNOLOGIES

First, let us reiterate Tim Berners-Lee's Linked Data principles:

1. Use URIs as names for things.
2. Use HTTP URIs so that people can look up those names.
3. When someone looks up a URI, provide useful information, using the standards (RDF, SPARQL).
4. Include links to other URIs so that they can discover more things.

To summarize, the Semantic Web relies on a stack of technologies, from URIs, through RDF for data, SPARQL for querying, and OWL and RDFS for ontologies, to the more immature although fast-developing areas of Provenance (with the Vocabulary of Interlinked Datasets, for example) and Trust. Although RDF can be serialized in XML, it is often easier to read and write in Turtle format, which exposes the triple structure more clearly. The learning point here, for those steeped in XML technologies or those familiar with the tabular form used to represent data in a GIS sitting on top of a relational database, is that RDF is more than just an XML format or a table structure: It is a *graph data model*, which allows knowledge to be structured much more flexibly, breaking away from both the relational and the document structure.

A second take-home message for Linked Data technologies is that we must be clear what exactly we are identifying with our URI: Is it a URI for a document describing the thing or the URI of the thing (the "resource") itself?

Third, we note that publishing Linked Data offers a more open and reusable alternative to the traditional REST (Representational State Transfer) API (Application Programming Interface) as a means to access data. Whereas with a REST API the structure of the data remains proprietary to each Web service, the RDF data model is standard across all data providers, making it much easier to integrate data from any Linked Data publisher.

Finally, as it has become apparent from our discussions in Chapter 8 onward, many of the technologies relating to Linked Data, particularly those relating to link discovery, authentication, and trust, as well as Linked Data user interfaces such as browsers, are still very immature. They are often the subject of university research projects and not at a stage where they can be reliably implemented in commercial, high-volume workflows, although things are changing rapidly.

The following are points to remember:

- RDF is a graph data model, most easily understandable using the Turtle format, which exposes the triple structure more clearly than the XML syntax.
- Separate the identification of a document or data about a resource from the URI of the resource itself so that your data is semantically accurate.
- The Linked Data approach is to publish using a standard data structure, namely RDF, which offers considerable advantages when it comes to data integration, compared with the proprietary data structure offered by a REST Web service API.
- Many Linked Data technologies, particularly those relating to link discovery, authentication, and trust, as well as Linked Data user interfaces such as browsers, are still very immature.

11.6 BENEFITS AND BUSINESS MODELS

There are four primary reasons for using Semantic Web technology, namely, for data integration; for data repurposing; for data collection, classification, and quality control; and finally for data publishing and discovery.

The majority of Linked Data is open and free at the point of use with varying licensing restrictions. This model is attractive to many governments where there is the need to make their data more accessible to the population. Much open Linked Data has also been published by individuals and voluntary groups such as GeoNames, where the driver is an altruistic desire to provide publicly available resources.

A number of commercial business models have been suggested for Linked Data publication: the subsidy model, increasing traffic to the publisher's site, advertising, certification, affiliation, service bundles and data aggregation, branding, and finally the direct payment model, either pay as you go or via subscription. This last case is most likely to occur as part of a freemium model, where some portion of the data is provided free and users upgrade to the paid version for enhanced data access.

Of all the business model options, we would argue that the freemium model is of the most interest to a commercial GI publisher since it can provide the openness required to improve discoverability while also protecting the value of the data; however, to date only the subsidy model has really been implemented.

The following are points to remember:

- Semantic Web technology brings benefits to data integration; data repurposing; data collection, classification, and quality control; and data publishing and discovery.
- The majority of data currently in circulation is published as open data by governments and voluntary groups.
- Thus far, only the subsidy business model has been used, although we would recommend a freemium model for GI publishers.

11.7   FUTURE DIRECTIONS

Finally, let us look into the future and make some suggestions for what may be to come at the crossroads between GI, Linked Data, and the Semantic Web. In doing so, though, bear in mind what was said by Niels Bohr, the great physicist: "Prediction is very difficult, especially about the future." To confirm the difficulty with prediction we write this book as new technologies are arising at a seemingly ever increasing rate. At the time of writing the industry has seen much interest in NOSQL (standing for Not Only SQL—not No SQL as is commonly thought). This technology has arisen as a result of an increasing need to process "Big Data", a terribly vague term referring to data that has at least one of the characteristics of complexity, rapidity of update, and/or size; these characteristics are often summarized as the three Vs of Variety, Velocity, and Volume. Here conventional SQL databases are seen to be inadequate and so new database types are emerging. Triplestores fall within the scope of NOSQL databases although for some reason the newer NOSQL technologies are seen by some as competing with Triplestores, RDF, and Linked Data. However the reality is that the technologies are complementary, and the confusion is more related to a misunderstanding as to what these technologies are trying to achieve and the niches that each tries to occupy. Indeed the confusion can be closely related to the misunderstanding that NOSQL means No SQL, not Not Only SQL. Not Only SQL summarizes the true situation very well: they are not intended to replace SQL databases but to co-exist with them. Linked Data fits within this world as a technology very well suited to handling complexity and data integration, it is not so strong at handling data that is updated or streamed at very high rates.

One thing that we are very certain of, and this book would be unnecessary if it were not true, is that the amount of GI will increase and its scope will broaden. It will do so as the Linked Data Web grows at an increasing rate, much as the document Web did.

We believe there will be a snowball effect in the emergence of commercial Linked Data. Companies seeing their competitors publishing Linked Data will realize that they also must have a presence on the Linked Data Web to drive traffic to their sites. Just as having a Web site is no longer a mere vanity project for companies, so will linking in to the Linked Data Web become a business essential as a way of disseminating data or as a fundamental publishing method. While early commercial Linked Data projects have been motivated by branding or a wish to try out the technology, primarily based on the subsidy business model, the subscription and pay-as-you-go models will emerge, probably through a freemium route. This in turn will necessitate development of better methods, it is hoped standards driven, to enable authentication and payment for Linked Data.

As we have pointed out, software tools to support link discovery and data reuse (particularly around the areas of evaluation of data quality, trust, provenance, and licensing) are still relatively immature. We predict that this situation will improve as tools emerge from university settings and become more robust and commercialized. There is a strong need for more analytical tools, especially with respect to data mining, and we also see a place for Linked Data Web analytics to track triple usage and provide publishers with more accurate information about which resources are the most valuable and frequently retrieved. Tools and standards will also develop to express GI as Linked Data more clearly, and the GeoSPARQL extensions of SPARQL will be seen in time as a fundamental part of the query language, just as geo extensions to SQL are now treated as just another capability of that language. We would like to see better tools for the end user, such as souped-up browsers that can handle geospatial queries, and we also expect to see Geospatial Linked Data more widely used within mashup applications.

As the Linked Data Web grows, so will ontologies and reusable micro-ontologies. These will probably not be sufficient in themselves to deal with the demands that are placed on them to process and connect data, so it is likely that there will be developments and increased use of rule-based languages such as RIF to complement OWL.

We should not assume that Semantic Web technologies have "solved" the data management and integration problem; it had not by any means been solved using traditional XML and database technologies, and it still remains the most thorny issue on the Semantic Web. However, what semantic technologies have provided is explicit methods to aid the resolution of this problem and opened it up on the Web, and we believe that shedding light on the dark corners of GI data integration can only make for a happier future.

11.8 CONCLUDING THOUGHTS

Geography and the Semantic Web share a common characteristic: They are both aids to data integration. Geography provides a means to connect information through shared location, the Semantic Web through shared identity. Together, they can begin to move data integration from an art and cottage industry to science and factory. Neither geography nor the Semantic Web approach is a cure-all. Much data has no natural geographic aspect, and some datatypes are not suitable for expression using Semantic Web technologies. But, there are always limitations with all things, so the important thing is to understand where and when they are applicable.

Our last piece of advice is true for any new or unfamiliar topic or technology: Start gently, start small, try to understand the underlying principles, experiment, and iterate to build on your successes. For geography, an important starting point is to properly understand identity and classification, that is, the nature of the things you are dealing with. To tackle the Semantic Web, it is important to gain an understanding of the open world assumption and then start by expressing some simple data in triple form. Once you gain confidence, you may wish to publish your triples as Linked Data, paying particular attention to openness, reuse, and descriptions of provenance and then try creating links to other data. From there, you may wish to experiment with more detailed description of this data by building ontologies that will further aid integration. This is the way that we propose you get to grips with Semantic Web technologies—Linked Data first, ontologies later—and it is reflected in the structure and order of the book.

However, once you have got to grips with the technologies, a more appropriate development workflow would be to start with the ontology. We strongly advise this approach as developing the ontology will provide a systematic framework overlying your Linked Data.

As you describe your GI (and indeed other data) ontologically and express it as Linked Data, you will inevitably encounter situations where what you want to say cannot be described and expressed as you wish. Here, our advice is to accept the limitations of Linked Data expression and develop means to manage them. We have indicated some ways to do this, but no book is able to provide a comprehensive coverage of all possible solutions to potential knowledge modeling problems. But, do not despair; you should be guided by the scope and purpose of what you are trying to do, coupled with an understanding of what semantic technologies *can do* to best manage these limitations. And, despite what the technologies cannot do, they do offer a fresh and powerful means to express GI, one that helps to make GI accessible to a much wider audience than before.

We conclude by voicing our hope that this book has been informative and helpful. With your newfound understanding of the nature of GI and the Semantic Web, we hope you are sufficiently confident to start experimenting with the publication of your GI as Linked Data and subsequently to express its meaning using ontologies. Enjoy the journey.

NOTE

1. Indeed, a characteristic of many geographic features is that they have either indistinct or unknown boundaries.

References

Adida, B., and Birbeck, M. 2008. RDFa primer—bridging the human and data webs. W3C recommendation. http://www.w3.org/TR/xhtml-rdfa-primer/.

Alexander, K., Cyganiak, R., Hausenblas, M., and Zhao, J. 2011. Describing linked datasets with the VoID Vocabulary. W3C Interest Group Note 03 March 2011. http://www.w3.org/TR/void/.

Auer, S., Dietzold, S., Lehmann, J., Hellmann, S., and Aumueller, D. 2009. Triplify—lightweight linked data publication from relational databases. In *Proceedings of WWW 2009*, Madrid, Spain.

Barrasa, J., and Gómez-Pérez, A. 2006. Upgrading relational legacy data to the semantic web. In *Proceedings of the 15th International Conference on the World Wide Web (WWW 2006)*, pp. 1069–1070, Edinburgh, UK, 23–26 May 2006.

Beckett, D. 2004. RDF/XML syntax specification (revised). W3C recommendation 10 February 2004. http://www.w3.org/TR/rdf-syntax-grammar/.

Beckett, D., and Berners-Lee, T. 2011. Turtle—Terse RDF Triple Language. W3C team submission 28 March 2011. http://www.w3.org/TeamSubmission/turtle/.

Berners-Lee, T. 1989. Information management: a proposal. http://www.w3.org/History/1989/proposal.html

Berners-Lee, T. 1996. Universal Resource Identifiers—axioms of Web architecture. http://www.w3.org/DesignIssues/Axioms.html

Berners-Lee, T. 1998a. Notation 3. http://www.w3.org/DesignIssues/Notation3

Berners-Lee, T. 1998b. Relational databases on the Semantic Web. http://www.w3.org/DesignIssues/RDB-RDF.html

Berners-Lee, T. 1998c. Semantic Web Roadmap. http://www.w3.org/DesignIssues/Semantic.html

Berners-Lee, T. 2006. Linked Data design issues. http://www.w3.org/DesignIssues/LinkedData.html. Accessed February 2012.

Berners-Lee, T., and Cailliau, R, 1990. WorldWideWeb: proposal for a hypertexts project. http://www.w3.org/Proposal.html. Accessed February 2012.

Berners-Lee, T., Hendler, J., and Lassila, O. 2001. The Semantic Web. *Scientific American* 284, 5, 28–37.

Bizer, C., and Cyganiak, R. 2007. The TriG syntax. http://www.wiwiss.fu-berlin.de/suhl/bizer/TriG/Spec/. Accessed May 2012.

Bizer, C., Jentzsch, A., and Cyganiak, R. 2011. State of the LOD cloud (August 2011). http://www4.wiwiss.fu-berlin.de/lodcloud/state/.

Boley, H., Paschke, A., Tabet, S., Grosof, B., Bassiliades, N., Governatori, G., Hirtle, D., Shafiq, O., and Machunik, M. 2011. Schema specification of RuleML 1.0. http://www.ruleml.org/spec

Bouquet, P. 2007. Contexts and ontologies in schema matching. In *Proceedings of the 2007 Workshop on Contexts and Ontology Representation and Reasoning*. Roskilde, Denmark, August 21, 2007.

Brickley, D., and Guha, R.V. 2004. RDF Vocabulary Description Language 1.0: RDF Schema. W3C recommendation 10 February 2004. http://www.w3.org/TR/rdf-schema/.

Brickley, D., and Miller, L. 2010. FOAF Vocabulary Specification 0.98 Namespace Document 9 August 2010. http://xmlns.com/foaf/spec/.

Brin, S., and Page, L. 1998. The anatomy of a large-scale hypertextual Web Search engine. In *Proceedings of the 7th International Conference on the World Wide Web (WWW)*, pp. 107–117, Brisbane, Australia. http://dbpubs.stanford.edu:8090/pub/1998-8

Broekstra, J., and Kampman, A. 2003 SeRQL a second generation RDF query language. http://www.w3.org/2001/sw/Europe/events/20031113-storage/positions/aduna.pdf

Carroll, J., Hayes, P., Bizer, C., and Sticker, P. 2005. Named graphs, provenance and trust. In *Proceedings of the 14th International Conference on the World Wide Web (WWW) 2005*, Chiba, Japan.

Choi, N., Song, I.Y., and Han, H. 2006. A survey on ontology mapping. *SIGMOD Record* 35 (September) 34–41.

Cobden, M., Black, J., Gibbins, N., Carr, L., and Shadbolt, N. 2011. A research agenda for linked closed dataset. In *Proceedings of the Second International Workshop on Consuming Linked Data (COLD2011)*, Bonn, Germany, 23 October, 2011. Edited by O. Hartig, A. Harth, and J. Sequeda.

Cohn, A.G., and Gotts, N.M. 1996. The "egg-yolk" representation of regions with indeterminate boundaries. In *Geographic Objects with Indeterminate Boundaries 2*, Burrough, P.A. and Frank, A. (eds.), Taylor & Francis, pp. 171–187.

Connolly, D., van Harmelen, F., Horrocks, I., McGuinness, D., Patel-Schneider, P., and Stein, L. 2001. DAML+OIL reference description. W3C Note 18 December 2001. http://www. w3.org/TR/daml+oil-reference.

Cregan, A., Schwitter, R., and Meyer, T. 2007. Sydney OWL syntax—towards a controlled natural language syntax for OWL 1.1. OWL Experiences and Directions Workshop, Innsbruck, Austria.

Cycorp. 2002. The syntax of CYCL. *Ontological Engineer's Handbook* v0.7, Chapter 2. http:// www.opencyc.org/.

Cyganiak, R., Delbru, R., and Tummarello, G. 2007. Semantic Web crawling: a sitemap extension. http://sw.deri.org/2007/07/sitemapextension/DERI, Galway.

Das, S., Sundara, S., and Cyganiak, R. 2011. R2RML: RDB to RDF mapping language. *W3C Working Draft 20 September 2011*. http://www.w3.org/TR/r2rml/.

Deakin, R. 2008. *Wildwood, a Journey through Trees*. Penguin, New York.

Dean, M., and Schreiber, G. 2004. OWL Web Ontology Language reference. W3C Recommendation 10 February 2004. http://www.w3.org/TR/owl-ref/.

De Nicola, A., Missikoff, M., and Navigli, R. 2009. A software engineering approach to ontology building. *Information Systems* 34, 258–275.

Dodds, L., and Davis, I. 2012. Linked data patterns. Accessed October 6, 2012 from http:// patterns.dataincubator.org/book/index.html

Du Charme, B. 2011. *Learning SPARQL*. O'Reilly, Sebastopol, CA.

Egenhofer, M. 1989. A formal definition of binary topological relationships. *Lecture Notes in Computer Science*, 367, 457–472.

Feigenbaum, L., and Prud'hommeaux, E. 2008. SPARQL by example: a tutorial. http://www. cambridgesemantics.com/2008/09/sparql-by-example/. Accessed 18 March 2012.

Fernández-López, M., Gómez-Pérez, A., and Jursito, N. 1997. METHONTOLOGY: from ontological art towards ontological engineering. In *Proceedings of the Spring Symposium on Ontological Engineering of AAAI*, pp. 33–40, Stanford University, Stanford, CA.

Fitzpatrick, B., Slatkin, B., and Atkins, M. 2010. PubSubHubbub Core 0.3 working draft. http://code.google.com/p/pubsubhubbub/. Accessed 20 May 2012.

Funk, A., Tablan, V., Bontcheva, K., Cunningham, H., Davis, B., and Handschuh, S. 2007. CLOnE: Controlled Language for Ontology Editing. *Lecture Notes in Computer Science*, 4825, 142–155.

Gangemi, A. 2005. Ontology design patterns for Semantic Web content. In *Proceedings of the Fourth International Semantic Web Conference*, M. Musen et al. (eds.), Springer, Berlin, 2005. pp. 262–276.

Gangemi, A., Guarino, N., Masolo, C., Oltramari, A., and Schneider, L. 2002. Sweetening ontologies with DOLCE Aldo knowledge engineering and knowledge management: ontologies and the Semantic Web. *Lecture Notes in Computer Science*, 2473, 223–233, DOI: 10.1007/3-540-45810-7_18.

Gearon, P., Passant, A., and Polleres, A. SPARQL 1.1 update. W3C Working Draft 05 January 2012. http://www.w3.org/TR/sparql11-update/.

Genesereth, M., and Fikes, R. 1992. Knowledge interchange format reference manual, Version 3.0, June. http://www-ksl.stanford.edu/knowledge-sharing/kif

Grant, J., and Beckett, D. 2004. RDF test cases. W3C Recommendation 10 February 2004. http://www.w3.org/TR/rdf-testcases/.

Haarslev, V., and Möller, R. 2001. RACER System Description. In *International Joint Conference on Automated Reasoning, IJCAR'2001*, R. Goré, A. Leitsch, and T. Nipkow (eds.), pp. 701–70518–23 June, Siena, Italy, Springer-Verlag, Berlin.

Hart, G., Johnson, M., and Dolbear, C. 2008. Developing a controlled natural language for authoring ontologies. *Lecture Notes in Computer Science*, 5021, 348–360.

Heath, T., and Bizer, C. 2011. *Linked Data: Evolving the Web into a Global Data Space*. Synthesis Lectures on the Semantic Web: Theory and Technology, 1:1, 1–136, Morgan & Claypool, San Francisco.

Hellmann, S., Auer, S., and Lehmann, J. 2009. Linkedgeodata—adding a spatial dimension to the web of data. In *Proceedings of the International Semantic Web Conference*, Chantilly, VA, October 25–29, 2009.

Hoekstra, R., Breuker, J., Di Bello, M., and Boer, A. 2007. The LKIF core ontology of basic legal concepts. In *Proceedings of the Workshop on Legal Ontologies and Artificial Intelligence Techniques LOAIT 2007*. Stanford University, Palo Alto, CA, June 4, 2007.

Hogan, A., Harth, A., Passant, A., Decker, S., and Polleres, A. 2010. Weaving the pedantic Web. In *Proceedings of the Linked Data on the Web Workshop (LDOW2010)*, Raleigh, NC, 27 April, CEUR Workshop Proceedings. CEUR-WS.org/Vol-628/ldow2010_paper04.pdf

Hogan, A., Harth, A., Umbrich, J., Kinsella, S., Polleres, A., and Decker, S. 2011. Searching and browsing Linked Data with SWSE: the Semantic Web search engine Web Semantics: science, services and agents on the World Wide Web. *Journal of Web Semantics* 9(4), 365–401.

Horridge, M., and Patel-Schneider, P.F. 2009. OWL 2 Web Ontology Language Manchester Syntax. W3C Working Group Note 27 October 2009. http://www.w3.org/TR/owl2-manchester-syntax/.

Horrocks, I., Patel-Schneider, P.F., Boley, H., Tabet, S., Grosof, B., and Dean, M. 2004. SWRL: A Semantic Web Rule Language combining OWL and RuleML. W3C member submission 21 May 2004. http://www.w3.org/Submission/SWRL/.

Huxhold, W. 1991. *An Introduction to Urban Geographic Information Systems (Spatial Information Systems*. Oxford University Press, New York.

International Organization for Standardization (ISO). *ISO 15836:2009 Information and Documentation—The Dublin Core Metadata Element Set*.

International Organization for Standardization (ISO). *ISO 19125-1:2004 Geographic Information—Simple Feature Access—Part 1: Common Architecture*. Accessed February 2012 from http://www.iso.org/iso/iso_catalogue/catalogue_tc/catalogue_detail.htm?csnumber=40114

International Organization for Standardization (ISO). *ISO 19128:2005. Geographic Information—Web Map Server Interface*.

International Organization for Standardization (ISO). *ISO 19136:2007. Geographic Information—Geography Markup Language (GML)*.

International Organization for Standardization (ISO). *ISO 19142:2010. Geographic Information—Web Feature Service*.

Kaljurand, K., and Fuchs, N.E. 2007. Verbalizing OWL in attempt to controlled English. OWL Experiences and Directions Workshop, Innsbruck, Austria.

Karvounarakis, G., Alexaki, S., Christophides, V., Plexousakis, D., and Scholl, M. 2002. RQL: A declarative query language for RDF. *Eleventh International World Wide Web Conference (WWW'02)*, Honolulu, Hawaii, 7–22 May.

Knublauch, H. 2011. SPIN—modeling vocabulary. W3C member submission 22 February 2011. http://spinrdf.org/spin.html. Accessed May 2012.

Longley, P., Goodchild, M., Maguire, D., and Rhind, D. 2001. *Geographic Information Systems and Science*. Wiley, New York.

Maala, M.Z., Delteil, A., and Azough, A. 2007. A conversation process from Flickr tags to RDF descriptions. In *Proceedings of the BIS 2007 Workshop on Social Aspects of the Web (SAW-2007)*, D. Flejter and M. Kowalkiewicz (eds.), Poznan, Poland, 27 April. CEUR Workshop Proceedings volume 245. http://ceur-ws.org/Vol-245/paper5.pdf

Manola, F., and Miller, E. 2004. RDF primer. W3C Recommendation 10 February 2004. http://www.w3.org/TR/rdf-primer/.

Miles, A., and Bechhofer, S. 2009. SKOS Simple Knowledge Organization System reference. W3C Recommendation 18 August 2009. http://www.w3.org/TR/skos-reference.

Mooney, P., and Corcoran, P. 2011. Annotating spatial features in OpenStreetMap. In *Proceedings of GISRUK 2011*, Portsmouth, England, April 2011. http://www.cs.nuim.ie/~pmooney/websitePapers/GISRUK2011_Peterv3-FinalSubmitted-MonFeb28th2011.pdf. Accessed February 2012.

Moreau, L., Clifford, B., Freire, J., Futrelle, J., Gil, Y., Groth, P., Kwasnikowska, N., Miles, S., Missier, P., Myers, J., Plale, B., Simmhan, Y., Stephan, E., and Van den Bussche, J. 2011. The Open Provenance Model core specification (v1.1). *Future Generation Computer Systems* 27(6), 743–756.

Motik, B., Parsia, B., and Patel-Schneider, P.F. 2009. OWL 2 Web Ontology Language: XML serialization. W3C Recommendation 27 October 2009. http://www.w3.org/TR/2009/REC-owl2-xml-serialization/.

Musen, M.A. 1988. Protégé ontology editor. http://protege.stanford.edu/Conceptual Models of Interactive Knowledge- Acquisition Tools Knowledge Acquisition, 1, 73–88, 1988.

Noy, N.F., Chugh, A., Liu, W., and Musen, M.A. 2006. *A Framework for Ontology Evolution in Collaborative Environments*, 5th International Semantic Web Conference, Athens, GA.

Noy, N.F., and McGuinness. D.L. 2001. *Ontology Development 101: A Guide to Creating Your First Ontology*. Stanford Knowledge Systems Laboratory Technical Report KSL-01-05 and Stanford Medical Informatics Technical Report SMI-2001-0880, March. http://protege.stanford.edu/.

Noy, N., and Rector, A. 2006. Defining *N*-ary relations on the Semantic Web. W3C Working Group Note 12 April 2006. http://www.w3.org/TR/swbp-n-aryRelations.

Noy, N., and Stuckenschmidt, H. 2005. Ontology alignment: an annotated bibliography. In *Proceedings of the Semantic Interoperability and Integration conference*. Kalfoglou Y., Schorlemmer, W.A., Sheth, A.P., Staab, S., and Uschold, M. (eds.). Publisher IBFI, Schloss Dagstuhl, Germany. Dagstuhl Seminar Proceedings volume 04391. 2005.

Open Geospatial Consortium. 2011. OGC Reference Model, OGC 08-062r7, v2.1 http://www.opengis.net/doc/orm/2.1

Palma, R., Haase, P., Corcho, O., and Gomez-Perez, A. 2009. Change representation for OWL 2 ontologies, *Proceedings of OWL: Experiences and Directions 2009 (OWLED 2009)*. Sixth International Workshop, Chantilly, VA, October 23–24, 2009.

Pease, A., Niles, I., and Li, J. 2002. *The Suggested Upper Merged Ontology: A Large Ontology for the Semantic Web and Its Applications*. AAAI Technical Report WS-02-11. http://www.aaai.org/Papers/Workshops/2002/WS-02-11/WS02-11-011.pdf.

Perry, M., and Herring, J., eds. 2011. GeoSPARQL—a geographic query language for RDF data. A proposal for an OGC Draft Candidate Standard. http://www.w3.org/2011/02/GeoSPARQL.pdf. Accessed February 2012.

Popitsch, N., and Haslhofer, B. 2010. DSNotify: handling broken links in the Web of Data. *WWW 2010*, Raleigh, NC, 26–30 April.

Prud'hommeaux, E., and Seaborne, A. 2008. SPARQL query language for RDF. W3C Recommendation 15 January 2008. http://www.w3.org/TR/rdf-sparql-query/.

Randell, D.A., Cui, Z., and Cohn, A.G. 1992. A spatial logic based on regions and connection. In *Proceedings of the 3rd International Conference on Knowledge Representation and Reasoning*, pp. 165–176. Morgan Kaufmann, San Mateo, CA

Rector, A.L., Rogers, J.E., and Pole, P. 1996. The GALEN High Level Ontology Conference: Medical Informatics in Europe 96—Human Facets in Information Technologies. In *Studies in Health Technology and Informatics*, Brender, J., Christensen, J.P., Scherrer, J.-R., and McNair, P. (eds.). Volume 34, pp. 174–178, 1996.

Rosenblatt, M. 1956. Remarks on some nonparametric estimates of a density function. *Annals of Mathematical Statistics* 27, 832–837.

Sauermann, L., and Cyganiak, R. Cool URIs for the Semantic Web. W3C Interest Group Note 03 December 2008. http://www.w3.org/TR/cooluris/.

Schreiber, G., Akkermans, H., Anjewierden, A., de Hoog, R., Shadbolt, N., Van de Velde, W., and Wielinga B. 1999. *Knowledge Engineering and Management: The CommonKADS Methodology*. MIT Press.

Schumm, S., and Kahn H. 1972. Experimental study of channel patterns. *Bulletin of the Geological Society of America* 83, 1755–1770.

Seaborne, A. 2004. RDQL—a query language for RDF. W3C member submission 9 January 2004. http://www.w3.org/Submission/RDQL/.

Sirin, E., Parsia, B., Grau, B.C., Kalyanpur, A., and Katz, Y. 2007. Pellet: a practical OWL-DL reasoner. *Web Semantics: Science, Services and Agents on the World Wide Web* 5(2), 51–53.

Smith, B., and Mark, D.M. 2003. Do mountains exist? Towards an ontology of landforms. *Environment and Planning B* 30(3), 411–427.

Smith, M.K., Welty, C., and McGuinness, D.L. 2004. OWL Web Ontology language guide. W3C Recommendation 10 February 2004. http://www.w3.org/TR/owl-guide/.

Stickler, P. 2005. CBD—Concise bounded description. W3C member submission 3 June 2005. http://www.w3.org/Submission/CBD/.

Stocker, M., and Sirin, E. 2009. PelletSpatial: a hybrid RCC-8 and RDF/OWL reasoning and query engine. OWL Experiences and Directions workshop. Chantilly, VA, October 23–24, 2009.

Story, H., and Corlosquet, S. 2011. WebID 1.0 Web Identification and Discovery. W3C Editor's Draft 12 December 2011. https://dvcs.w3.org/hg/WebID/raw-file/tip/spec/index-respec.html.

World Wide Web Consortium (W3C). 2009. OWL 2 Web Ontology Language document overview. W3C Recommendation 27 October 2009, W3C OWL Working Group (eds.). http://www.w3.org/TR/owl2-overview/.

Yu, A. 2011. *Developer's Guide to the Semantic Web*. Springer-Verlag, New York.

Appendix A

OWL Species

OWL (Web Ontology Language) is not a simple language; it has distinct dialects or species and subspecies. Figure A.1 nicely shows this linguistic complexity. However, things are not quite as bad as they may first appear. There have been two versions of OWL standardized: OWL 1 became a World Wide Web Consortium (W3C) Recommendation in 2004 (Dean and Schreiber, 2004) and consists of three "species": OWL Full, OWL DL, and OWL Lite. This was followed by OWL 2, which was standardized in 2009 (W3C 2009); OWL 2 is an evolution of OWL 1. As well as OWL 2 DL and OWL 2 Full, there are three sublanguages of OWL 2 DL offered: OWL 2 QL, OWL 2 EL, and OWL 2 RL. So, what are the differences and when should each be used?

OWL DL[1] is the most widely used version of OWL by far; indeed, often references to "OWL" are actually referring to OWL DL. All OWL languages are based on first-order logic; the *DL* in OWL DL stands for Description Logic, which is the name given to the subset of first-order logic that OWL DL uses. OWL DL was designed to balance expressivity (being able to express complex knowledge structures) against computational completeness (any statement that can be made in the OWL DL language is either true in the ontology or false) and decidability (if the truth or falsehood of a new statement can be determined based only on the set of statements provided in the ontology). It also took into account which practical reasoning algorithms were available at the time the language was designed to compute the logical consequences of the ontology statements. That is, OWL DL was designed to be able to say as much as possible and make complicated, detailed statements about the world while knowing that the reasoning calculations would actually come back with an answer within a finite time. Every ontology language has to take a stance on this issue of balancing expressivity against efficiency of reasoning: how much can be said versus how quickly, whether one or every answer can be reached by the reasoner, or indeed whether an answer can be reached at all.

OWL Lite is a sublanguage and a subset of OWL DL that was specified for the benefit of early tool builders who wanted to get started. Now, it is rarely used, and you are unlikely to come across any ontologies that specify themselves to be "OWL Lite." (On a side note, many ontologies will not need to use every single type of expression that is possible in OWL DL, but their "expressivity" (which level of complexity of expression they use) will in practice be determined by the content of the domain, not by the mathematics of the various types of logic. Hence, it is nearly impossible to find a domain that fits precisely into the OWL Lite subset.) OWL Full offers more expressiveness than OWL DL by forgoing any computational guarantees; that is, no answer may ever be reached. It uses different semantics than

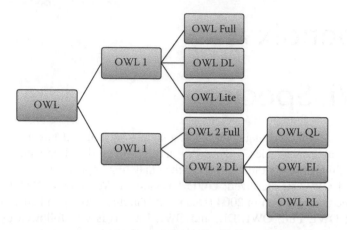

FIGURE A.1   The OWL family.

OWL DL; for example, it allows a class both to be treated as a set of individuals, as usual, and to be treated as an individual itself. It was designed to be compatible with RDF (Resource Description Framework) Schema, but there are no practical OWL Full reasoners available, and it is rarely used in real-world situations.

In OWL 2, OWL 2 DL and OWL 2 Full species offer additional constructs to their version 1 namesakes. OWL 2 offers an additional three sublanguages, known as profiles. Since OWL 2 is a very expressive language, it can be difficult to implement and work with, so the profiles, which are all more restrictive than OWL DL, offer an easier approach to OWL. The profiles make different trade-offs between certain computation or implementation benefits against various aspects of OWL's expressivity. OWL 2 EL was designed primarily for large biohealth ontologies. It is suitable for any ontology that, like the biohealth ones, requires complex descriptions of structures, for example, describing body parts in terms of what parts they contain. Any domain that includes complex structures, such as scientific domains, system configurations, or product inventories, would suit OWL 2 EL. OWL 2 EL offers the performance guarantee of reasoning within a finite, polynomial time and so should be chosen when the ontology is very large.

The second sublanguage, OWL 2 QL, is targeted at databases, with the *QL* standing for Query Language, as this profile can be implemented by rewriting queries into a standard query language for relational databases, such as SQL. It is meant to be integrated with relational databases and benefit from their robust implementations and scale. It is suitable for representing database schemas as it can be used to describe Entity Relationship and UML diagrams and for integrating schemas via query rewriting. Given this type of use, OWL 2 QL is suitable for lightweight ontologies with large numbers of individuals, stored in a database. The reasoning belongs to a different complexity class than OWL 2 EL, but one whose problems are also considered feasible to solve in polynomial time.

OWL 2 RL can be used when you want to add rules to RDF data, with the *RL* referring to Rule Language. It is useful for OWL 2 applications that have a greater need for efficiency than expressivity and for RDF applications that need to be enhanced with some additional expressivity from OWL. This means that it is particularly suitable for applications that use lightweight ontologies to organize large numbers of individuals or if you need to operate on RDF triple data.

NOTE

1. For simplicity, we use OWL Full and OWL DL to denote both versions of these languages and only use OWL 2 Full or OWL 2 DL if there is a difference that requires highlighting. OWL 1 is a subset of OWL 2; every OWL 1 ontology is a valid OWL 2 ontology.

REFERENCE

Dean, M., and Schreiber, G. OWL Web Ontology Language reference. W3C Recommendation 10 February 2004. http://www.w3.org/TR/owl-ref/.

Appendix B

OWL Constructs: Manchester Syntax and Rabbit

Construct	Rabbit	Manchester OWL Syntax
Ontology Declaration		Ontology: <http://mereamaps.gov.me/topo> <http://mereamaps.gov.me/topo-v1>
Import	Use http://purl.org/dc/elements/1.1/	Import: <http://purl.org/dc/elements/1.1/>
Annotation Property		AnnotationProperty: dc:rights
Class Declaration	Stream is a Concept.	Class: Stream SubClassOf: owl:Thing
Object Property Declaration	"is contained in" is a Relationship that takes a concept as an object.	ObjectProperty: isContainedIn
Datatype Property Declaration	"has name" is a Relationship that takes a value as an object.	DatatypeProperty: hasName
Individual Declaration	England is an individual.	Individual: england
Sub Class Of	Every Bourne is a kind of Stream.	Class:Bourne SubClassOf: Stream
Equivalent Classes	Petrol Station and Gas Station are equivalent.	Class: PetrolStation EquivalentTo: GasStation
Disjoint Classes	River and Floodplain are mutually exclusive.	Class: River DisjointWith: Floodplain
Some Values From	Every Pub sells Beer.	Class: Pub SubClassOf: sells some Beer
All Values From	Every Basin is connected to only a Channel or a Pipe or nothing.	Class: Basin SubClassOf: isConnectedTo only (Channel or Pipe)
Has Value	Every Loch is located in Scotland.	Class: Loch SubClassOf: isLocatedIn value Scotland
Object Union	Every Mission has purpose one or more of Christian Worship or Charitable Activities.	Class: Mission SubClassOf: hasPurpose some (ChristianWorship or CharitablePurpose)
Object Intersection	Every School has a Building that has purpose Education.	Class: School SubClassOf: Building and hasPurpose some Education
Object Complement Of	No Backwater has a Current.	Class: Backwater ObjectComplementOf: hasCurrent some Current

Construct	Rabbit	Manchester OWL Syntax
Object One Of	Every UK Country is exactly one of England or Wales or Northern Ireland or Scotland.	Class: UKCountry ObjectOneOf: england, wales, northern_ireland, Scotland
Object Exact Cardinality	Every River Stretch has part exactly one Channel.	Class: RiverStretch SubClassOf: hasPart exactly 1 Channel
Object Min Cardinality	Every River Stretch has part at least two Banks.	Class: RiverStretch SubClassOf: hasPart min 2 Bank
Object Max Cardinality	Every River Stretch has part at most two confluences.	Class: RiverStretch SubClassOf: hasPart max 2 Confluence
Class Assertion	England is a Country.	Individual: england Type: Country
Same Individual	Portsmouth and Pompey are the same thing.	Individual: portsmouth SameAs: pompey
Different Individual	England and Scotland are different things.	Individual: England DifferentFrom: scotland
Object Property Assertion	Portsmouth is located in Hampshire.	Individual: portsmouth Facts: isLocatedIn hampshire
Negative Object Property Assertion	Portsmouth is not located in Scotland.	Individual: portsmouth NegativeObjectPropertyAssertion: isLocatedIn scotland
Object Property Range	The relation "is capital city of" can only have a Country as an object.	ObjectProperty: isCapitalCityOf Range: Country
Object Property Domain	The "is capital city of" relationship can only have a Capital City as a subject.	ObjectProperty: isCapitalCityOf Domain: CapitalCity
Datatype Property Range	The relation "has name" can only have a String as a value.	DataProperty: haName Range: xsd:string
Equivalent Object Properties	The relationships "is inside" and "is within" are equivalent.	ObjectProperty: isInside EquivalentTo: isWithin

continued

Construct	Rabbit	Manchester OWL Syntax
Disjoint Object Properties	The relationships "contains" and "is contained in" are mutually exclusive.	`ObjectProperty: contains` `DisjointWith: isContainedIn`
Inverse Object Properties	The relationship "contains" is the complement of "is contained in".	`ObjectProperty: contains` `InverseOf: isContainedIn`
Sub Object Property	The relationship "flows into" is a special type of the relationship "flows".	`ObjectProperty: flowsInto` `SubPropertyOf: flows`
Symmetric Object Property	The relationship "is adjacent to" is symmetric.	`ObjectProperty: isAdjacentTo` `Characteristics:Symmetric`
Asymmetric Object Property	The relationship "is larger than" is asymmetric.	`ObjectProperty: isLargerThan` `Characteristics: Asymmetric`
Reflexive Object Property	The relationship "is near to" is reflexive.	`ObjectProperty: isNearTo` `Characteristics: Reflexive`
Irreflexive Object Property	The relationship "flows into" is irreflexive.	`ObjectProperty: flowsInto` `Characteristics: Irreflexive`
Transitive Object Property	The relationship "is part of" is transitive.	`ObjectProperty: isPartOf` `Characteristics: Transitive`
Functional Object Property	The relationship "has postcode" can only refer to one thing.	`ObjectProperty: hasPostcode` `Characteristics: Functional`

Inverse Functional Object Property	The relationship "is assigned to" can only have one subject.	`ObjectProperty: isAssignedTo`
		`Characteristics:`
		`  InverseFunctional`

Sub Property Chain	Everything that has a Part that contains something will also contain that thing.	`ObjectProperty: contain`
		`subPropertyChain: hasPart o`
		`  contain`

Index

A

Abox, *see* Assertion Box (ABox)
"Abstract," 14
abstraction, 105
accuracy, *see also* Preciseness
    correctness, 132
    imprecision, 208–210
    incorrect markup, 20
    reuse, 122–123
    using Linked Data, 157
ACE, *see* Attempto Controlled English (ACE)
AceView, 180227
addresses
    data integration, 39
    structure, 32
    textual representations, 31–33
adjacency information, 114
"adjacent," 189
admin prefixes, 141
advertising
    data publishing and discovery, 18
    models, 138
affiliation models, 138–139
Agents, 127
aggregation models, 139
Agricultural Production example, 196, 197
agriculture, GIS purpose, 37
aliases, 147
Allegro Graph, 130
Altby example, 211–212
Altova SemanticWorks, 181
amateur communities, 47–49
Amazon
    data sources with APIs, 120
    historical developments, 42
ambiguity
    classification, 30
    mereology, 101
AND, 173
"and," 173–174
angle brackets
    tags, 20
    Turtle, 75
anonymity, 70
Apache Jena Framework, 130
API, *see* Application Programming Interface (API)
application ontologies, 184
Application Programming Interface (API)
    accessing Web information, 10

    data sources with, 120–121
    publishing Linked Data, 10
    Search Monkey, 21
    technologies, 230
"a" predicate, 76
Arable land example, 173
archaeological find example, 27
ARC/INFO, 38–39
ArcView, 39
arithmetical analysis, 59
Artifacts, 127
n-ary relation, 200–202
Ash Fleet Farm examples, *see* Building
          geographic ontologies; Linked Data,
          organizing GI as
ASK keyword, 143–144
Assertion Box (ABox), 168, 170
assignment, order of, 89
asymmetric property, 176
Atom syndication format
    data sources with APIs, 121
    recent trends, 21
Attempto Controlled English (ACE), 166
attributes, OGC features, 52–53
aunt/niece example, 207
authentication
    certification, 138
    publishing Linked Data, 127–128
authoring, tools for, 179–181
automatic link discovery and creation, 153
axioms, 166, 171

B

Bakery example, 215–216
Basic Geo Vocabulary, 68, 111
Bayesian networks, 153
BBC, 137
becks, 190
beer examples, 69–70, 148, 169
bee waggle dance, 37
"belongs to," 171
benefits, Semantic Web, 16–19, 231
Berners-Lee, Tim
    designing and applying URIs, 89
    five-star rating system, 128
    Linked Data principles, 107
    semantic spam, 158
    Web historical developments, 40

Milton Keynes UK
Ingram Content Group UK Ltd.
UKHW040445071024
449327UK00020B/1014